Praise for *Across a Waking Land*

'This book is an open-hearted, humble and gritty adventure with a deep love of nature at its heart. It takes courage to walk alone across the land, and it takes passionate curiosity to find out what lies behind the green fields of Britain. Roger Morgan-Grenville possesses these qualities, they shine through every page. *Across a Waking Land* is a masterful interweaving of personal endeavour with the big conservation challenges of Britain today. I loved it.'

MARY COLWELL, author of *Curlew Moon*

'Beautifully captures the essence of a British spring'

Birdwatcher Magazine

'Prescient, perceptive and powerful: an articulate and thoughtful account of nature's increasingly fragile state, experienced through an advancing spring.'

TIM BIRKHEAD, author of *Birds and Us*

Praise for *Taking Stock*

'Funny, insightful and hugely informative ... a charming book'
Daily Mail

'Tremendous ... We all need to take stock, and this is the ideal starting point. I learnt a lot from this book and laughed a lot too.'
ROSAMUND YOUNG, author of *The Secret Life of Cows*

'Stylishly locates the importance of the cow in absolutely everything from finance ... to future food. [A] first-prize rosette for this paean to the wonderful cow, Man's other best friend.'
JOHN LEWIS-STEMPEL, *Country Life*

'A lyrical and evocative book'
Daily Express

'No cow could ever hope for a better appreciation of its truly unique worth.'
BETTY FUSSELL, author of *Raising Steaks: The Life and Times of American Beef*

'An epic story told with warmth, wit and humanity. Will make us feel differently about these long-suffering animals.'
GRAHAM HARVEY, author of *Grass-Fed Nation*

Praise for *Shearwater*

'This charming and impassioned book meanders, shearwater-like, across a lifetime and a world, a rich tribute to an extraordinary bird drawn through tender memoir and dauntless travel.'
   HORATIO CLARE, author of *A Single Swallow* and *Heavy Light*

'*Shearwater* is sheer delight, a luminous portrait of a magical seabird which spans the watery globe'
   Daily Mail

'This is wonderful: written with light and love. A tonic for these times.'
   STEPHEN RUTT, author of *The Seafarers: A Journey Among Birds*

'What I love about Roger Morgan-Grenville's writing is the sheer humanness of it … Bravo – a truly lovely book.'
   MARY COLWELL, author of *Curlew Moon*

'*Shearwater* is a delightful and informative account of a lifelong passion for seabirds, as the author travels around the globe in pursuit of these enigmatic creatures.'
   STEPHEN MOSS, naturalist and author of *The Swallow: A Biography*

'A captivating mix of memoir, travel and ornithological obsession … A book not just for seabirders or island-addicts, but for all who have ever gazed longingly out to sea and pondered vast possibilities and connections.'
   BBC Wildlife magazine

'A beautiful mix of memoir and natural history … entirely infectious.'
   Scottish Field

Praise for *Liquid Gold*

'A great book. Painstakingly researched, but humorous, sensitive and full of wisdom. I'm on the verge of getting some bees as a consequence of reading the book.'
    CHRIS STEWART, author of *Driving Over Lemons*

'A light-hearted account of midlife, a yearning for adventure, the plight of bees, the quest for "liquid gold" and, above all, friendship.'
    Sunday Telegraph

'*Liquid Gold* is a book that ignites joy and warmth'
    MARY COLWELL, author *of Curlew Moon*

'Beekeeping builds from lark to revelation in this carefully observed story of midlife friendship. Filled with humour and surprising insight, *Liquid Gold* is as richly rewarding as its namesake. Highly recommended.'
    THOR HANSON, author of *Buzz: The Nature and Necessity of Bees*

'Behind the self-deprecating humour, Morgan-Grenville's childlike passion for beekeeping lights up every page. His bees are a conduit to a connection with nature that lends fresh meaning to his life. His bee-keeping, meanwhile, proves both a means of escape from the grim state of the world and a positive way of doing something about it. We could probably all do with some of that.'
    DIXE WILLS, *BBC Countryfile magazine*

'Both humorous and emotionally affecting ... Morgan-Grenville's wry and thoughtful tale demonstrates why an item many take for granted should, in fact, be regarded as liquid gold.'
    Publishers Weekly

By the same author
Across a Waking Land (Icon, 2023)
Taking Stock: A Journey Among Cows
(Icon, 2022)
Shearwater: A Bird, An Ocean, And a Long Way Home
(Icon, 2021)
Liquid Gold: Bees and the Pursuit of Midlife Honey
(Icon, 2020)
Unlimited Overs (Quiller, 2019)
Not Out of the Woods (Bikeshed Books, 2018)
Not Out First Ball (Benefactum Press, 2013)

THE RESTLESS COAST

THE
RESTLESS
COAST

A Journey around
the Edge of Britain

Roger Morgan-Grenville

ICON

For Tom and Alex, with much love and in the hope that you will discover for yourselves the full range of wonders of your ancestral coastline a few decades earlier than your father managed.

Published in the UK and USA in 2025 by
Icon Books Ltd, Omnibus Business Centre,
39–41 North Road, London N7 9DP
email: info@iconbooks.com
www.iconbooks.com

ISBN: 978-183773-144-2
ebook: 978-183773-146-6

Text copyright © 2025 Roger Morgan-Grenville

The author has asserted his moral rights.

No part of this book may be reproduced in any form, or by any means, without prior permission in writing from the publisher.

Typesetting by SJmagic DESIGN SERVICES, India

Printed and bound in the UK

Appointed GPSR EU Representative: Easy Access System Europe Oü, 16879218
Address: Mustamäe tee 50, 10621, Tallinn, Estonia
Contact Details: gpsr.requests@easproject.com, +358 40 500 3575

CONTENTS

Prologue: An End, and a Beginning — xvii

PART 1: WATER

1 The Blue Anthropocene Coast — 3

PART 2: SOUTHERLY

2 A Salmon Dilemma: Sutherland and the Outer Isles — 17
3 The Honey Pot Island: The Sea Eagles of Mull — 39
4 Restoring the Commons: First Aid for Seabeds in Argyll and Beyond — 59
5 Art, Activism, and a brief introduction to Nature Conflict — 75
6 The Lost Kingdom of Cantre'r Gwaelod: Cardigan Bay — 91

PART 3: EASTERLY

7 The Sea as Bringer: The South West Coast — 109
8 A Bay of Sewage and Sex-changing Fish — 129
9 Three feet high, and Rising — 149

PART 4: NORTHERLY

10	A Wild Goose Chase: In Search for Abundance	173
11	The Shrinking Caravan Coast: East Yorkshire	191
12	Dead Crabs on a Troubled Shoreline: North Yorkshire and Teesside	211
13	Black-eyed Gannets and the Power of Hope: Northumberland	235
14	Salar: North-east Scotland	247

Epilogue: Dunnet Head	255
Acknowledgements	261
Bibliography	265

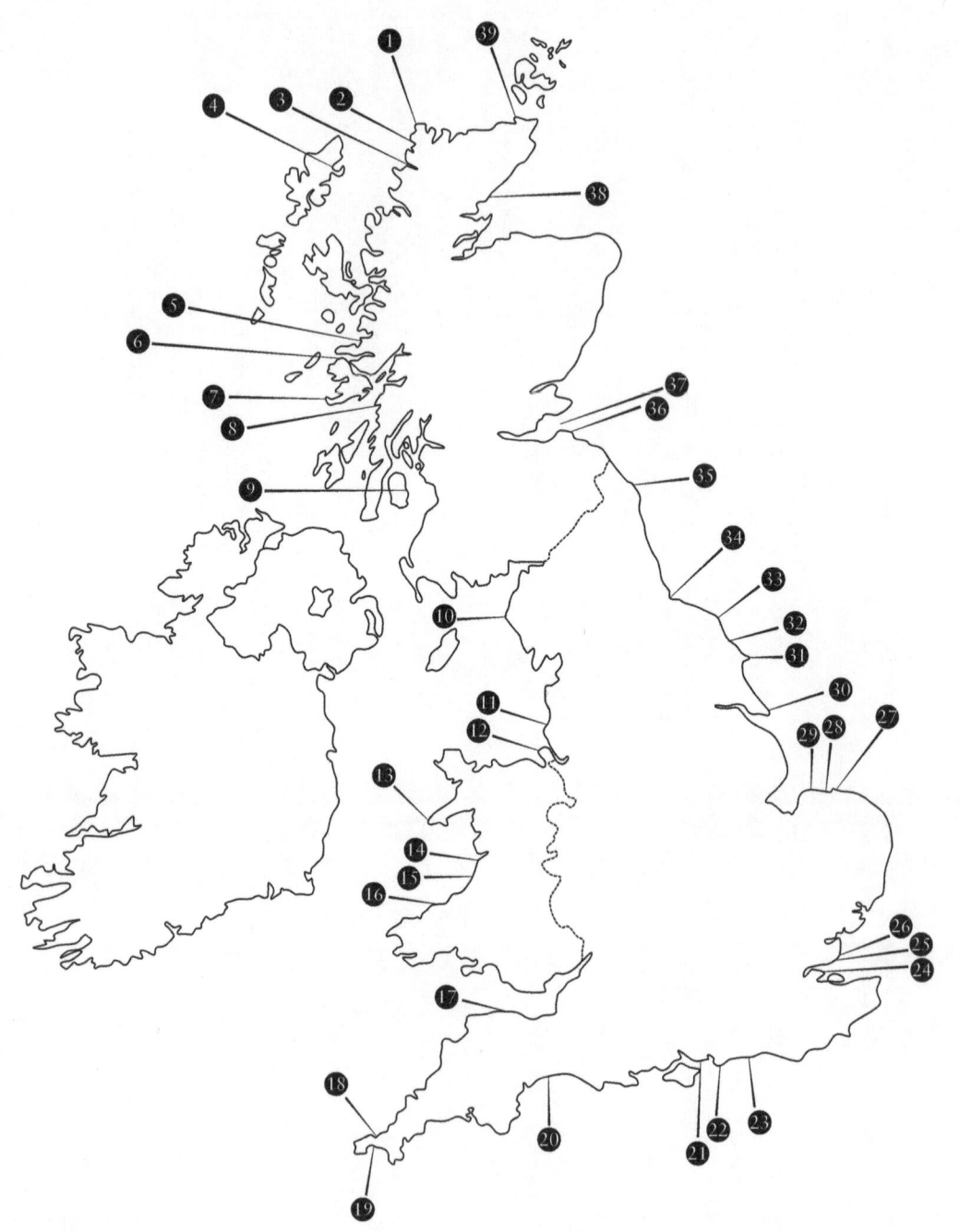

KEYS

1. Cape Wrath
2. Sandwood Bay
3. Badcall Bay
4. Stornoway
5. Ardtoe
6. Tobermory
7. Iona
8. Ardfern
9. Lamlash Bay
10. Ravenglass
11. Crosby
12. Hoylake
13. Bardsey Island
14. Fairbourne
15. Aberystwyth
16. Newquay
17. Minehead
18. Hayle Beach
19. Penzance
20. Lyme Regis
21. Portsmouth
22. Langstone Harbour
23. Littlehampton
24. Dartford
25. London
26. Broomway
27. Holkham
28. Titchwell
29. Wild Ken Hill
30. Spurn Head
31. Flamborough Head
32. Scarborough
33. Whitby
34. Hartlepool
35. Seahouses
36. North Berwick
37. Isle of May
38. Brora
39. Dunnet Head

PROLOGUE: AN END, AND A BEGINNING

'The sea-shore is a sort of neutral ground, a most advantageous point from which to contemplate the world.'

Henry David Thoreau

This story found life in the shadow of a lighthouse.

It was late spring, and I had just finished walking for 55 days up the 1,000-mile spine of Britain from the Solent on the south coast to the north-west corner of Scotland; I was resting against a wall at Cape Wrath, eating a damp cheese and pickle sandwich and watching the tireless fulmars.[1] Tireless fulmars are good things to watch when you, yourself, are tired; their rapid rising and falling among the updraughts and eddies of the cliff air creates their own energy, a tiny part of which it is easy to imagine transferring to you. Oddly, the feeling of accomplishment that might reasonably have been expected to stir in someone middle-aged after such a long walk had been instead hijacked by the sense of a something not finished, of there somehow being more to come. For a while, I just put it down to the faint grief

---

[1] A story told in *Across a Waking Land*. Icon Books, 2023.

that inevitably attends the end of any long journey, but the sight of a little yacht far below, butting its brave way through the swell and out towards the Orkney Islands, and then a separate one, a fishing boat, turning south towards Kinlochbervie, said something entirely different. It talked of new journeys, and of this moment being only halfway at best. It talked of the coast.

Contrary to what most of us might logically assume, that the 'wrath' in Cape Wrath refers to the local anger of the stormy weather and violent seas around it, it actually derives from the old Norse word for 'turning point'. And, as it was for the Vikings, resetting the sails of their longships for the southerly voyage down the indented western shores of 'Bretland', so it became for me. Deep down, so deep that it took me a good six months to come to fully acknowledge it myself, it was calling for me to one day turn on my heel and follow that fishing boat down the coast as far as she and her inshore fellows would lead me. All the way round the south coast and back to the north coast, maybe. This tattered, watery edge of my otherwise tinder dry world hid a fresh world of cliffs and coves, ferries and fishing villages, salt marshes and salmon farms, all awaiting discovery. As Robinson Crusoe found out, to understand your island a bit better, you first have to understand what surrounds it.

It is all but pointless to expect a journey to provide definitive answers to specific questions, and anyway, that's not really the point of them. I wasn't looking for comprehensive answers, so much as a better understanding of how my coastline worked, what threatened it and what people were doing about it. By seeking the answer, we assume that there is a clear 'right' and a 'wrong', which there generally isn't, and we over-narrow the frames of reference of the trip, and thus stop being inquisitive. And if we stop being inquisitive, we might as well just hurry to the destination as quickly as possible and hope that the weather is good and the views agreeable. Better instead just to travel and let the miles tell you whatever story they will. Then, if you walk enough miles, you'll know more than you did when you started, and care more, and want to do more.

A while after my return home from Cape Wrath and back in downland Sussex, I stuck a large map of Britain up on the wall of the room where I do my writing, because that's how and where adventures begin, with clean maps and a blank canvas. Lines from various sharpies started to appear on its laminated surface over the next few weeks until there were thirteen of them spaced around the whole length of the coast,[2] with small orange, blue and yellow sticky circles placed along them at varying intervals denoting things that I particularly wanted to see. Thirteen different coastal foot journeys, then, over the course of twelve months, to allow me to know a little better the thin liminal ribbon of sand, rock, mud, marsh and grass that separates our island from the ocean beyond, and the people who live and work there. From the seafood that we harvest from its estuaries and bays to the power we harness from above its offshore shelves, from the warming and rising of its seas to the steady erosion of its moveable coastline, there are things happening wherever you look, and always there are people to tell their stories.

A hundred million years ago, the spot on the earth's surface where I am now standing was the southern tip of Greenland; a hundred million more, and it was Greenland's northern tip, another hundred, this was water somewhere off Alaska, and finally, half a billion years ago, I would be treading water right in the middle of the largest body of sea the planet has ever seen. To watch a time lapse reconstruction of the movement of the tectonic plates beneath us is to see the movement of land masses on an unthinkable scale, and you are either the sort of person who is delighted or overawed

---

[2] This is the eternal unanswered question: how do you measure a coastline? The Ordnance Survey figure for the mainland of Britain is 17,819 km, and that is good enough for me.

by the fact that we are moving still, and at the same rate. I know my journey cannot be through deep time, but it is good for my humility to understand the cosmic insignificance in space and time represented by one middle-aged man walking on and off for a year or so around the coastline of his island home.

What followed was a journey of discovery, of anger, of inspiration and of hope. It involved almost every coastal county in Britain, just under 2,000 miles of walking and hundreds of bus, train and ferry trips. I even rediscovered hitch-hiking after a gap of 40 years. Joyfully, cars were rarely involved, as they generally make lousy companions on solo linear walks. Besides, you don't need £4 and someone's permission to park your feet.

I would start, I decided, on World Ocean Day at the very lighthouse where I had finished the last journey.[3] And then I would just see what happened.

---

[3] 8 June 2023

PART 1
WATER

THE BLUE ANTHROPOCENE COAST

1

An Introduction

> 'No water, no life. No blue, no green.'
>
> Sylvia Earle, US Marine Biologist

In 1972, as part of the final Apollo mission, one of the most iconic images of all time was captured. We call it the Blue Marble. This snapshot of the world, taken from outer space, is now seen as a pivotal moment in helping people understand the sheer fragility and isolation of our home. But humanity's understanding of blue as a distinct colour arrived relatively late; surprisingly, given their seemingly endless coastline, the ancient Greeks had no word for blue, and no real concept of it. For that matter, and for a long time, nor did anyone else.

It was not until 431 CE, when the Catholic church got round to colour-coding the saints and just happened to dress the Virgin Mary in a blue robe, that blue finally became accepted as a proper colour, just as red, green, black and white had been accepted for millennia.[1] It was so costly to produce artificially that, even

---

[1] Interestingly, it was a further thousand years before the colour orange was described in isolation.

then, it would then take another 1,000 years for mainstream art to fully accept and utilise it. And yet blue is the most prevalent colour by far of our planet, because three-quarters of its surface is water. The blue of this water is the dominant shade that we reflect back when seen from space, but it is also a tricky colour, almost ephemeral at times. Because we perceive blue at a short wavelength, it scatters before it arrives, almost as if it gets lost. Blue is another world.

Within the blue ocean's depths are thought to be contained around 1.3 billion cubic kilometres of salt water or, to put it another way, 95 per cent of all the available habitat on our planet.[2] The sea's presence just happens to be what makes life on our planet possible, and from it all life on earth originally arose. Within its depths, countless tiny phytoplankton produce, as a biproduct of photosynthesis, much of the oxygen that will eventually allow us to breathe. At the same time, the sun heats the sea water, which evaporates and condenses to form clouds which then become the precipitation that waters our crops and slakes our thirsts; a trillion tons of it are evaporated from the surface of the oceans, each and every day.[3] Moreover, the largest proportion of protein that we as a species eat still comes from the seafood that we harvest from or grow in its depths. 'The ocean,' as one scientist phrased it, 'is an engine for converting sunlight into movement and life and complexity, before the universe reclaims the loan.'[4]

We may be land creatures, but we are utterly reliant on the salt water around us. Indeed, according to one anthropological theory, by surgeon and amateur evolutionary biologist Peter Rhys Evans,

---

[2] *The High Seas*. Olive Heffernan. Profile Books. 2024.
[3] From *Blue Machine*, by Helen Czerski. A powerful and beautifully written scientific guide to the hidden world of the world's oceans.
[4] Ibid.

our evolution may have been rather more associated with coastal waters than we think, and that it is not for nothing that we became hairless apes with variable skin colours, strange inner ear bones and the unique ability to hold our breath.[5] For a land mammal, our unusual taste for a protein source from a different element to our own suggests, at the very least, a deep cultural connection with what lies offshore. Either way, the sea defines us, and water is, according to the most famous polymath of all, 'the driving force of all nature'.[6]

Although the sea is one continuous body of water, we have traditionally divided it for convenience into four principal oceans. On the swirling, north-eastern edge of one, the Atlantic, our little archipelago is an almost insignificant outpost, no more than 0.05 per cent of the surface area of the earth. And, depending on where you draw the line on the definition of 'island', our principal island is surrounded by around 6,300 smaller ones, as well as the islets, skerries and rocks that make up what is the twelfth longest national coastline in the world. When you multiply the length of our coast by its great tidal range, we probably have more intertidal area than any other country on earth. Just about every drop of rain that falls on our fields, and each breath of wind that cools us, is influenced by our being an island, and generally as a result of what goes on over the western horizon. And perhaps the most influential of all those things going on is the 60-mile-wide horizontal column of warm water that is, for now, winding its way towards us at about five miles per hour across the Atlantic; because of the Gulf Stream, our climate is perpetually mild, and our winters are more like Memphis than Moscow.

---

[5] *The Waterside Ape: An Alternative Account of Human Evolution*. Peter Rhys-Evans. CRC Press. 2020.

[6] Leonardo da Vinci

Our reputation for talking incessantly about the weather is often overstated,[7] but it is rooted in the waves around us. Because of the shape of our country, no one in it can be more than 45 miles from tidal water, and only that much if they happen to be in the village of Coton in the Elms, in Derbyshire. There is almost no part of our island story that does not have roots in the sea around us. We are, to paraphrase one of our more distinguished post-war politicians,[8] simply 'a lump of coal surrounded by fish'. But, before you get to the fish, you get to the coast.

We tend to see the coastline between our land and that ocean as something permanent, yet it is anything but, not least because the whole island group has been inching itself northwards on its long voyage from what is now Antarctica for at least the last 600 million years. Even today, it is not just Brexit that is pulling us away from Europe; in a metaphor rich with political irony, our landmass is heading west towards America, although not quite as fast as America is heading away from us. The great storms that batter our western edges are obvious agents of erosion and change, as is the water that our river systems bring to the sea, but there are far more powerful and subtle dynamic forces at work, too. Most of us are blissfully unaware that, while one part of our main island (the north and western bit) is still springing gently ever higher in isostatic reaction to the weight of ice it lost fifteen millennia ago,[9] the opposite, south-eastern, part is actually sinking into its own compacting clay at the rate of about five centimetres a century.[10] When you add this shrinking phenomenon to a sea level that is set to rise as a

---

[7] Anecdotally, the Scandinavians do it far more than the British.

[8] Aneurin (Nye) Bevan MP (1897–1960). What he then went on to say was that 'only an organizing genius could produce a shortage of coal and fish at the same time'.

[9] Although the MCCIP 2020 report card indicated that, for the first time, sea level rise was outstripping the vertical land movement.

[10] Compare this with Indonesia's capital, Jakarta, where some areas have reportedly subsided 16 feet in the last 25 years (Bloomberg).

direct result of the warming climate by a not so gradual centimetre every three years, you can accurately predict the slow inundation of a swathe of the east coast, and the reappearance of salt water in the streets of cities as far apart as Gloucester and Glasgow. Cardiff and London are assessed to be third and seventh respectively on the list of large cities on earth that are most vulnerable to the rising sea.[11] The Thames Barrier, originally designed to be raised a couple of times a year, is in reality raised three times that amount already. On the Holderness Plain in Yorkshire, erosion sees to it that roads that once went somewhere now come to a yawning halt on clifftops above a sea that is chewing up around two metres of land a year.

There is much that we still don't know, but we can say for certain that our island is still changing shape and doing so at a geological warp speed; so much so that, according to a 2022 report from the Environment Agency, not far off a quarter of a million of our homes will need to be abandoned in the next thirty years, and double that number of people will be displaced. This is not entirely surprising. After all, we weren't even an island 10,000 years ago and, since then, the water around us has risen, in a process that predates any fixed notion of climate change, by over 30 metres, which is well over half the height of Nelson's Column in Trafalgar Square. This sinking feeling is not new alarmism, so much as a relentless natural process way beyond our power to influence. Our planet is older than we care to think, and our part in it no more than a slight clearing of the throat in its long story: if we think of the history of the Earth in terms of a 24-hour clock, the dinosaur extinction event took place 21 minutes ago, and all our own written history is compacted into the last tenth of a second.[12]

Far from being a single entity, our long coastline plays host to a wide variety of different habitats, each one a sensitive miracle of

---

[11] Earth.org
[12] *Otherlands*. Thomas Halliday

natural diversity and variety. Mudflats that arise from eons of settling silt brought by rivers from the land further inshore, and are teeming with worms, snails and bivalves; saltmarshes that host up to 150 unique species of invertebrates and help prevent our villages from flooding; machair dune grassland with its corncrakes and twites; lagoons, rocks, shingle and the towering weather-battered cliffs and stacks of the north that give a home only to lichens and dense colonies of seabirds. We have even created habitats of our own, such as our coastal cities, or the Wallasea Island Wild Coast project in Essex, where the three million tons of London earth that has been extracted for Crossrail has created a new 700-hectare wildlife reserve of mudflat, saltmarsh and grassland, butting out into the sea. And yet there is a remorseless shrinking of available intertidal habitat going on, caused by coastal squeeze, a process through which mudflats and saltmarsh are lost to erosion and higher sea levels, but where the presence of coastal defences are preventing new habitats developing.[13]

Out in the sea and below our horizon, even the most bullish trawlerman would struggle to argue that the last century had been a good one for the fish numbers in our waters. Compare and contrast the record-breaking year of 1911, when over a *billion* herring were decanted onto the docksides of East Anglia, with the situation now, when 'there are only two or three local fishermen working the sea there', and where 'the one remaining smokehouse processes herrings caught by Norwegians to be exported to Northern Europe'.[14] Unbelievably, our island nation still manages to be a net importer of fish, with 270,000 tons more coming in than we ship out.[15] As for herring, so for cod, salmon and sole: we eat more than it is in the gift of the ocean to provide us with. And, as for fish, so for invertebrates,

---

[13] *Coastal Adaptation to Benefit Wildlife.* Ausden et al. British Wildlife. Dec 2023.

[14] *Silver Shoals.* Charles Rangely-Wilson

[15] UK Sea Fisheries statistics 2019

mammals and birds. The kindest that could be said is that we have not been good custodians of the sea that we were given dominion over.

And imagine that, in that same sea, twice a day, every day, a giant body of salt water,[16] slow and irresistible, arrives at the Scilly Isles, off the west coast of Cornwall, and splits in two. One part rolls up the western coast, turns east at the top of Scotland and then heads down the eastern shore until, somewhere off the Thames Estuary, it meets with the other half, which has come up the English Channel. Then it turns around and retreats again the way it came. We may see it creeping inexorably up harbour walls in Cornish fishing villages, racing across the mud flats of Morecambe Bay or insinuating itself deftly into the finger-like creeks of the North Norfolk seaboard but, wherever we do, we will see it doing this with a greater range and energy than almost anywhere else in the world.[17] Then, a few hours later, we note again its retreat, carrying in its eddies and backwaters tiny traces of physical witness to our inland human impact, before it returns six hours later to start the adventure all over again. We call this hydraulic process the *tide*, and from its actions and bounty stem many of the thousands of ongoing biological processes and food chains that deliver to us the extraordinary plant, insect, fish, bird and mammal life that delights and sustains us. As Canute demonstrated to his sycophantic courtiers a thousand years ago, these are the living breaths of the ocean which even a king is powerless to hold back.[18] Even industrial man. Even us.

---

[16] The salt which makes up on average 3 per cent of sea water comes from a variety of sources, including from eroding rocks, and from vents and salt domes on the seabed (National Ocean Service). Only rainwater lacks any salt at all.

[17] The tidal range in the Severn Estuary is a full fifteen metres, second only in the world to the Bay of Fundy in Canada.

[18] Anyone who has experienced the joy of watching Knots foraging on, and almost against, the incoming tide, will readily understand why their Latin name is *calidris canutus*.

Often, because so much of what happens on the edgelands between land and sea is hidden from us by the dark water, we hardly know about our coast, or seem to care much about it. Maybe this is because our predisposition is to fear the unfamiliar element around us, and therefore consciously not want to discover too much about it, and risk disturbing the brooding magic. And it is a delicious fear, a fear that is woven into the very culture and metaphors of our grounded lives: 'we look to dry land, try to remain grounded, and we hope for plain sailing in the days to come'.[19] It is a fear of whirlpools and sea monsters, predators and poisoned barbs, a fear of the unseen five miles of mysterious danger below us in the unlikely event that we find ourselves treading water in the middle of the ocean. But it is also high adventure. Wherever we are on the coast, the proximity of the ever-changing sea constantly reminds us of what we are not, which is in control: around 250 us around these islands meet our end in its insistent, salty embrace each year,[20] and it is not for nothing that the ghosts of over 37,000 shipwrecks punctuate our coastline.[21]

These days, apart from those who make their homes and their living there, the coastal settlements sometimes seem to have become little more than places for the rest of us to go and lie on the sand, hold conferences and stag weekends and visit heritage museums. You hear the words 'used to be' a good deal in British seaside towns. It is as if we have taken those saucy postcards, piers and ice cream vans as the seaside's true *leitmotif*, and built our expectations of it from them. In stark contrast to many other coastal nations, wealth and enterprise have leaked away from the coast and been largely replaced by old people,

---

[19] *The Draw of the Sea*. Wyl Menmuir. 2022
[20] www.outdoorswimmingsociety.com
[21] Historic England.

low wages and poverty, other than the very best houses, which tend to be owned by faraway money. Granted, the more energetic of us occasionally tramp by foot from settlement to settlement along its wonderful coastal paths, but the vast majority will arrive and leave by car and train, that is if they haven't already flown off to lie in the sun by some other country's ocean. If we are not careful, we can start to see the seaside more as a mere source of entertainment and heritage than the life force it truly is, more a museum of the heady days of sexual licence and experimentation than a functioning modern community with the need for a future. As important as anything, the people who live on that coast, and who still strive to make a living around it, need to be able to fashion sustainable futures that involve more than a few temporary seasonal jobs in seaside pubs, or vulnerable jobs in a declining fishing industry.

From our own point of view, our role as islanders makes us bold and independently minded; seen from our neighbours on occasion, it makes us moody and treacherous. Whatever shade of that is true, it is also what our lives happen to be built on. And, in whatever fashion those lives happen to be built, the coast almost always exhales a sense of adventure and anticipation: from pirate stories to fish and chips, there is something for everyone.

For good or ill, it is we humans who are increasingly shaping that coast around the folds and pleats of the way we live. This chapter is called 'the Anthropocene Coast' in informal acknowledgement of this unofficial geological era in which we are now said to be living, which is itself derived from the Greek words for 'man' and 'new'.[22] It reflects the idea that the eight billion of

---

[22] In March 2024 scientists rejected the idea that there is now enough evidence to suggest that nuclear weapons testing, fossil fuel burning, and overexploitation have pushed us into this new era, even though they have often voted on candidate sites to represent Anthropocene's birthplace in case they eventually agree that there is.

us are now cumulatively such influential residents of the planet that we as a single species are starting to change its very future geology, through our contribution to ecosystem and climate change, pollution and species loss, let alone the cleanliness and health of its surface. If you were to narrow the effects of this influence down to our own tiny corner of the world, you would see it everywhere: in the poisoned shellfish and in the phosphates draining out of our agricultural rivers; in the deep-water ports and the crumbling man-made sea defences; in the depleted seagrass meadows and the barren seabeds below our salmon farms. Perhaps our biggest challenge is that, somewhere out there behind both the headlines and the headlands, our coastal waters are warming at an alarming rate, and so are the rivers that feed into them.

And yet we also need to remind ourselves of the good we do, and try to do, too, the beauty we have created and the sanctuary that we have started to provide in, for example, our much-argued over Marine Protection Areas and No-Take Zones. If we start to see it all as a lost cause, then how can we possibly expect our children to get out there and finish anything we have started? We have brought back the white-tailed sea eagle to its skies, the blue-finned tuna to its waters, and we have helped the rare natterjack toad to cling on for a new tomorrow. We farm sheep on Ramsey Island just to give the beleaguered chough a continuing toehold and we have restored dynamic dunescapes in North Devon that have brought back flowers not seen for half a century. Through rat eradication programmes, we have brought burrow-breeding seabirds back to islands from St Agnes to the Shiants. Humpback whales have recently been seen in Liverpool Bay for the first time since 1938, and otters regularly swim up the Mersey. We should celebrate the rivers such as that one and the Thames, which are cleaner now than they have been since before the Industrial Revolution, and let these little advances inform us all how much more we could achieve, and how quickly, if we have the mind to. It is as if the coast, that ribbon of delight stitched for ever around our lives, has only now become a live and exciting subject, something that we have the power to alter for the good.

Finally, we need to remember that the coast should be, and in recent times has been, a place for the visitor to breathe in uncomplicated joy. It may be that in the re-finding of that joy we discover some of the answers as to how to allow it to thrive in the future.

PART 2
SOUTHERLY

A SALMON DILEMMA: SUTHERLAND AND THE OUTER ISLES

2

Early June

> 'The aim of argument should not be victory, but progress'
> Karl Popper

A handful of miles from Cape Wrath, at the north-west tip of Great Britain, lies Sandwood Bay. It is a stretch of sand and dunescape of almost cliched beauty and remoteness.

Words and pictures generally fall short in describing it to others, so it is perhaps best just to go there yourself once in your lifetime, and to sit alone for an hour or so on one of the giant machair dunes behind it, drink it all in without overthinking it, and just let it thread into you whatever memories and emotions you happen to be receptive to. Anyway, if you have come from anywhere but Sutherland, it has taken you days to arrive here, so you should probably give it the time it deserves. Your attention might be drawn to the Shepherd (*Am Buachaille*), an unfeasibly tall and isolated sea stack out to the south, or to the ochre colour of the sand, or the rippling blue wildness stretching away up to Cape Wrath. It really doesn't matter. If you allow it to, and especially if you put the phone and the camera back into your pack where they belong, it is communicating directly with a part of you that is beyond thought, the part that admits to being awed. Breathe deep. In a fast-moving and fragile world that demands our complicit 24-hour attention and anxiety, it is uncomplicated wonder, and it is a nature cure. It seems as permanent as the stars.

But on that hot afternoon in June, I start to learn that this permanence is perhaps illusory. I have walked to Sandwood over the unforgiving moorland from Cape Wrath, and it has taken me some time to notice that something is adrift. I have been here before a few times, which means that there is an expectation loosely based on those memories, but it is an expectation that goes unfulfilled. Things that are missing are inevitably harder to spot than the things that are in front of you. For sure, there is much to delight: a little rock sculpture on the sand, the pink of the thrift and the yellow of the trefoil, the colourful tents of some wild campers above the dunes, and an old arthritic man walking his equally old arthritic spaniel along the tideline. There is the sultry song of the waves washing onto the sand, and a distant fishing boat that butts its way through the choppier offshore seas to its home at Kinlochbervie. High above, there are the contrails of a jet on its polar route to America, a journey that may be completed in not much more time than it has taken me to walk here from Cape Wrath. Then it dawns on me: the thing that is missing is birds. Where I might have expected raucous gulls and kittiwakes at the lighthouse, some foraging fulmars off the low cliffs, bustling cormorants at the bay and – a little further out – the brilliant white torpedo shapes of diving gannets, there is only the quiet of emptiness, the last knockings of a silent season. There are none anywhere to be seen. Gone also are the soaring flight songs of the lapwings from the moor, the restless piping of the oystercatchers from the sea rocks, the skylark. When the seabirds stop crying, it is as if the summer has lost its voice.

Away from the windswept ocean greyness for which their colouring has largely evolved, seabirds are easy to spot, being generally large and dominantly countershaded with white or black, and often operating in noisy groups. If they are around, you will almost certainly see or hear them. But if they aren't, you are slowly drawn into an ecology lesson, which leads inexorably to questions about our stewardship of this planet. Seabirds are good indicators, which means that through their shifting populations and changing behaviours, we can learn much about their entire ecosystems, and what we might have done to them. From the 12 million tons of plastics that find their way each year into our oceans,[1] to the rapidly warming sea around our island; from the depleted fish stocks,[2] to the man-made diseases that decimate their colonies from time to time, seabirds call plaintively to us to change our ways.[3] Sixty years ago, 1.5 billion of them made this call; these days, it is nearer 500 million. Extrapolate that into the middle distance, and you arrive in silence sooner rather than later. As an island of infinite variety, we are still rich in breeding seabirds – we have about 80 per cent of the world's Manx shearwaters and 56 per cent of its gannets[4] – but their continued presence here is not a given.

Of course, even as I heed the silence, I know in my heart that they will be back, that this is just a freak moment in time, a foreshadowing of the worst of avian flu, a coming together of both temporary and permanent absences to create a momentarily silent sky. Yet it comes as an uncomfortable foretaste, for silence is not in the nature of the coast. To any naturalist, silence calls out louder than the noisiest kittiwake colony.

---

[1] Surfers against Sewage.
[2] On 1 April 2024, sand eel fishing was finally banned English and Scottish waters, an initiative that DEFRA concluded would lead to a 7 per cent rise in UK seabird numbers within 10 years.
[3] Scottish beaches have some of the highest recorded levels of beach litter with 919 items per 100 metres, 94 per cent of which is plastic. *Marine Litter: An Assessment of sources, controls and progress in Scottish Seas*. Mcleod. 2024. Paper for Environmental Standards Scotland. (Not published yet.)
[4] Joint Nature Conservation Committee.

And perhaps no silence articulates more precisely the scale of our nature challenge than that of the gulls, and no place better illustrates this at the start of my journey than the ribbon of coast between our land and the wild blue yonder across the horizon.

A good proportion of the environmental problems that we face start with the food decisions we make or that are made for us. To satisfy our national hunger (the hunger from which 60 million of us will each try to eat 35 tons of food in our lifetimes, and waste a third of it), we need to have access to a vast amount of supplies. For this, half a million of us work in a national food industry in which about 55 per cent of produce is home grown,[5] and the rest imported from all around the world. Our 200,000 farms cover just under three-quarters of our land and are by far the most influential factor in how much nature will be around for our children to enjoy. Most of us are uninterested in the massive production and supply chains that lead to our dinner plates, and surprisingly relaxed about who we give our food money to. Other things divert us more easily. Phones, for example, and holidays abroad. A cynic might say that a country that delegates its national food policy to a handful of powerful supermarkets gets roughly what it deserves – a reliable supply of cheap, not very good food, sourced in an efficient but not very fair way. In politics, food and the environment always seem to come in a distant tenth to the imperative of being re-elected, a phenomenon that, in Scotland, often smells of fish.

That night, I treat myself to salmon for the first time in many years. My body says that I need some protein, and I am in the heartland of farmed salmon, so the food miles are theoretically close to zero. Besides, I used to love it before I gave it up.

---

[5] Food and Drink Federation.

Salmon farming, in principle, is much like any livestock farming, in that you breed a captive population of stock as fast as practically possible on its way to the human food chain. The captive stock, in the case of salmon, is in a process of endless selection for fast growth, and is contained in a netted pen (generally around 25–50 metres deep and 50 metres or so in diameter), which is located within tidal water that, theoretically, cleans the waste away every few hours. Generally fed on pellets containing a mix of vegetable and fish oil, the salmon grow to their marketable weight of around 4 kilos in somewhere under 2 years.

While there is evidence that humans have been farming fish in China for close to 5,000 years, salmon farming is a relatively new industry in Britain. From its humble beginnings on a small farm in Loch Ailort, Inverness-shire, in 1971, the extraordinary growth of aquaculture in Scotland has mirrored in an accidental way the equally extraordinary decline of wild fish out in the ocean. The predictability and safety involved in farming protein, as opposed to hunting it, makes a great deal of sense, on the face of it, and we have been doing it for 10,000 years. In the five decades since 1971, salmon farming has become huge business, stretching up Scotland's west coast like a 400-mile long necklace, from Arran in the south to the Shetland Islands in the north: 1 farm has become 213,[6] 14 tons has become 205,000,[7] and a handful of jobs has grown into around 2,000.[8] At roughly the same time, 1,220 tons of rod-caught wild Atlantic salmon in 1983 has become about a tenth of that;[9] indeed, in December 2023, the once plentiful Atlantic Salmon

---

[6] Scottish Government Fish Farm Production Survey 2021

[7] Ibid. There is talk of this rising to 500,000 tons.

[8] There is little agreement between the industry and its critics as to how many jobs are really involved, and how local they actually are. My estimate of 2,000, plus 8–10,000 in the wider supply chain, is not scientific.

[9] *Sea Change*, Richard Girling.

joined 157,190 other creatures on the IUCN Red List,[10] informally an extinction watch. Farmed salmon, who contribute £760 million to the Scottish economy each year,[11] are now the UK's biggest food export and are a highly nutritious and lowish carbon food source.[12] And it's not just salmon: aquaculture as an industry in Scotland also includes rainbow trout, oysters, mussels and scallops, not to mention freshwater hatcheries, across maybe 400 different sites.[13] These days, there is also a growing industry emanating from the harvesting of seaweed, that extraordinary and prolific coastal survivor that has evolved to thrive both in the wet and the dry, and in salt water and fresh. By any standards, aquaculture is a remarkable financial success story.

It is equally remarkable, then, just how viscerally salmon farming has come to be opposed, with criticism ranging from shocking welfare to antibiotic overuse, and seabed damage to the infection of wild salmon with sea lice. Indeed, there is a positive alphabet soup of pressure groups working on the matter, to the extent that occasionally calls to mind *Life of Brian*'s People's Front of Judea sketch.[14] 'Salmon farming has lost its social licence, for some reason,' said one industry figure to me. 'Well under a quarter of the people in Scotland actually oppose it, but the activists have become a very powerful and well-coordinated lobby.' To be fair, they probably need to be powerful

---

[10] International Union for Conservation of Nature.

[11] *Fish Farmer* magazine. Also a figure widely disputed by many.

[12] Or not. With feed shipped in from other continents and 40,000 tons or more of farmed salmon leaving Heathrow by plane each year to foreign markets, the overall carbon claim can look a bit flimsy. The good news is that a Faroese company, Hiddenfjord, has developed a way of shipping fresh salmon long distances without using aircraft; others might just follow.

[13] These days, London's Billingsgate Market sells more farmed fish than wild. (*Blue Machine*, Helen Czerski.)

[14] Ocean Rebellion, Sea Shepherd, Wild Fish, Scottish Salmon Watch, Salmon and Trout Conservation Scotland, Animal Equality and Viva, to name but a few.

and well-coordinated these days: most of the small players in this industry have either closed or been absorbed into the large ones, whose far-reaching interests are keenly defended by their industry body, Salmon Scotland, or Salmon Norway, as critics often refer to it. There is nothing wrong per se in foreign ownership of businesses, but the extent to which Norwegian companies control Scottish farmed salmon is remarkable, including by far the largest of them, MOWI.

An influential campaign called Off the Table is asking chefs and restaurants to pledge not to serve farmed salmon to their guests.[15] Meanwhile, in October 2023, the campaigning group WildFish submitted a letter to the Competition and Markets Authority over the industry's claims to be sustainable, which they dispute, and the industry's plan to change its protected name from 'Scottish Farmed Salmon' to 'Scottish Salmon', which they consider misleading.[16] And yet, in comparison, the broiler chicken industry, which produces 37,000 jobs and almost 1.1 billion birds in often highly questionable and polluting conditions, seems almost to get away without comment. Ditto pigs and often cattle.

It is partly to try to find an answer to the question 'should I be eating this stuff at all?' that I am walking down the West Sutherland coastline, home to a raft of these offshore farms. After all, the United Nations Food and Agriculture Organisation (FAO) has told us that, to protect the land, we need to derive more protein from the sea which, in default of anything else, means more farms. We humans eat 180 million tons of seafood a year, which makes fish still by far the main source of protein for the world's population.[17]

---

[15] At the time of writing, subscribers to the scheme included over 120 restaurants worldwide, including 67 in the UK and 40 in Iceland.

[16] Interestingly, part of the industry's defence of removing the word 'Farmed' from 'Scottish Salmon' is that no one can buy wild salmon any more, so their product is, de facto, the only salmon in town.

[17] *The Salt Roads*, David Goodlad.

For a while, I stare at the pink salmon steak on my white plate, and idly wonder at the basic strangeness of its life; how one of the world's great cold water migratory predators has been reduced to swimming around a small, circular tank of warmish seawater with a couple of hundred thousand competitors, and where one big farm can hold more fish than the entire global population of wild Atlantic salmon.[18] A fish that, when wild, is genetically matched to one specific river, must be at the very least confused by a circular pen. A fish that in order to conform with market expectations might have both carotenoids and oily fish added to the feed, for colouring and Omega 3 respectively, is not necessarily consistent with the natural beauty of the marketing images or, indeed, with nutritional excellence. A fish for whom the deformities of an over-rapid growth are commonplace. You are, as they say, what you eat eats, and not everything the fish on my plate ingested was free of chemicals. And it is probably a good idea to avoid deluding yourself that the fish on your plate has had what is generally understood to be a good life. Out there, in the vast oceanic food system, maybe; but in here, no. Not at all.

People don't generally think too much about fish welfare, but according to the critics, they should. Sea lice are an ever-present problem within the packed pens, with infestations tormenting the fish and, worse, crossing over to wild salmon who are even less able to deal with it. Occasional sea lice are part and parcel of a salmon's life, but when this rises to four or more per fish, as it does frequently, there is a major impact on the young salmon's welfare. More than 10 million salmon died pre-harvest on Scottish farms in both 2022 and 2023,[19] out of a total production of around 40 million. Recently, this has been added to by the arrival of a plague of microscopic jellyfish, hydrozoans,

---

[18] *Net Loss: the high price of salmon farming*. Article by Mark Kurlansky in *Guardian*, 15 Sep 2020.

[19] *The Economist* Report, July 2024.

who cause huge damage to gill health, which in 2022 nearly doubled reported pre-harvest mortalities from 8.5 to 14.9 million salmon,[20] often over 30 per cent of the total. If there is no solution found to this issue (which may or may not be exacerbated by the warming oceans), it could have existential consequences for the industry, never mind the fish. Don Staniford, a campaigner who has been a thorn in the side of the industry for years, points out that if ramblers saw one in four sheep or cows dead in a field, a similar percentage to the 2022 figure, they would be horrified and action would be taken.[21] The problem, he continues, is that what goes on in the fish pens is largely hidden and inaccessible. Tavish Scott, the CEO of the industry body Salmon Scotland, counters with: 'Wild Atlantic salmon have a survival rate of only around 1–2 per cent, compared to 85 per cent for a farmed salmon.' Keep in mind that the industry means for that 200,000 tons of salmon biomass already out there to double in the next ten years, with half of the increase already in the scoping and planning process. If you were searching for a slowdown, you won't find it here.

It would take an imaginative soul to think that life for a farmed salmon is good, but nothing is simple. For example, the Norwegian company MOWI[22] has recently been given a welfare award[23] by none other than Compassion in World Farming (CIWF). This is intriguing, given that pre-slaughter weekly mortality rates in MOWI's Scottish operation have often run at more than 10 per cent.[24] It seems at first glance rather like giving the Spanish Inquisition a

---

[20] Scottish Fish Health Inspectorate figures.
[21] Salmon deaths on Scottish Fish Farms. Article in *Guardian*, 15 Jan 2023.
[22] I spent the best part of a year trying to speak to MOWI to get their side of the story. Their continued silence explains better than I can why I am unable to tell it.
[23] Special Recognition Award at CIWF's Good Farm Animal Welfare Awards, 2022.
[24] MOWI Scotland: *Sea Lice and Mortality Reporting*. August 2023. (Refers to Bagh Dail Nan Cean Farm.)

prize for ecumenical endeavour, but then the world of research is always full of surprises to the open mind.

A few days later, I catch up briefly with Lou Valducci, CIWF's Head of Food Business, and suggest to her that a company whose intensively farmed fish had been secretly filmed being eaten alive by sea lice seemed, on the face of it, an unlikely recipient of a welfare award from a world-leading animal charity. In turn, she gives me a compelling argument for seeking compromise and incremental improvement.

'The actual award was for one small bit of their business,' she says, 'the new technique of non-percussive stunning that they now use to slaughter fish, and which is said to be much better for the fish in every respect.

'Our mission is a bit more complicated than just campaigning for a ban, so we accept that we're not going to get rid of it in the short term. Part of our work is wrapped up in future policy making to make that aim happen, but a large chunk consists of interventions in the here and now, so that farmers who are trying to get incrementally better, and are investing in getting better, don't constantly find the goal posts moving around them. Encouragement must play as big a part as criticism.

'In 2019, we put what we call "the Fish Ask" before the industry, and offered to work with companies who accepted our broad aims. That "ask" included a raft of measures across the whole industry, including how the fish were slaughtered. MOWI signed up; they developed this technique; we rewarded them for it. Simple.'

She has a point. In an industry that is apparently here to stay, it makes sense to work with it to make it better, in this case on welfare grounds. As veteran campaigner John Aitchison put it to me a few weeks later: 'Whatever damaging thing you want to do, you just need to do a less damaging version of it.'

Undeterred, I decide to have salmon again for dinner the following night, this time in Scourie, and after another twenty-mile walk around Loch Inchard and Loch Laxford. That makes twice in five years, but then this is research, and I intend to be open-minded. Conveniently so, for it happens to be delicious.

I stare at the fish on my plate as I did the previous evening, but this time I wonder about the effect that its short life has had on the environment around it.

The issue of pollution from salmon farms derives, of course, from having a very large number of fast-growing cold-water fish in a relatively small, warm and static location. They produce a large amount of high-nutrient waste, which sinks to the bottom, are fed medication, which leaks out with the tide, and they are often treated with pesticides and antibiotics. (One of aquaculture's ugly secrets is that, in the period from 2014 to 2021, when antibiotic use went down in every other land and water farm type by at least 50 per cent, in salmon farming it actually went up by 168 per cent.)[25] The question of what they eat is also key: composite pellets with varying degrees of wheat, soy and fish, of which much of the latter comes from Peruvian anchovies, themselves not only in sharp decline, but also no longer available as a local food source for humans. It is a sobering thought that around 30 per cent of all fish caught in the ocean are fed back to other fish, who in turn are being reared for humans to eat. The area of seabed immediately below a farm, known as the benthic layer, is particularly susceptible to de-naturing from the effects of all that fish poo: a relatively low threshold of provable life is required by the regulator, and the penalties for not achieving even that are small.

Until recently, half a million tons of threatened sand eels were hauled out of our waters each year, much of it to be pulped and fed back to the salmon in those shimmering tanks,[26] an extravagance

---

[25] Vet antimicrobial resistance surveillance report, 2021.
[26] In March 2023, the UK and Scottish governments announced measures to ban sand eel fishing in its waters, which came into force a year later.

that goes some way to explaining why the beleaguered puffin is on the BTO's red list.[27] Fish farming is also acknowledged by the regulator as likely to be adversely affecting wild salmon populations;[28] it is a matter of controversy, not least because the east coast salmon rivers, where there are no fish farms, are doing as badly as the west. However, with warming rivers, faster spates, more seals, the effects of hydro power and a thousand further challenges,[29] it is probably forgivable to resent any avoidable contribution to the extinction of a wild fish that has been feeding us for 48,000 years, and from whom arises a deep and rich seam of our own cultural story. To make things worse, the wild sea trout have suffered even more than the salmon, as they tend to hang around in the estuary, rather than migrate.

Critics point to all of this and try to warn us that warming seas are only going to exacerbate things, but the rest of us continue to march into our supermarkets and demand our twin pack of salmon steaks at £4.95.[30] Whatever else is going to trouble the farmed salmon industry in the near future, it will not be lack of demand. And sadly, whatever else is going to trouble the consumer, history does not suggest that it is going to be a sudden mass concern for welfare and the environment, at least not any time soon.

Unsurprisingly, though, the industry is now having to take these problems more seriously than its success might otherwise persuade it to. Using cleaner fish like wrasse to feast on the sea lice ('green washing', say the critics; it is pretty ineffective, and you still have to catch the wrasse, feed them and kill them when you harvest the salmon). They are starting to experiment with closed containment farms and even onshore facilities ('counter-productive', say the

---

[27] British Trust for Ornithology 'Into the Red', 2022.
[28] Scottish Environmental Protection Agency (SEPA) report.
[29] One of which is no longer netting, which was banned in Scotland from 2015.
[30] Tesco online. 2 boneless salmon fillets. November 2024.

critics about the latter, as you just take a low carbon food and make it a high carbon one). There is talk of taking farms ever further out to sea, but that, in a world of increasing extreme weather events, could lead to major storm damage and uncontrolled break-outs. Nothing is easy. Lurking in the background is the decision of the US Food and Drug Administration (FDA) back in 2015 that genetically modified salmon was fit for human consumption, and with no requirement to include the GM information on the label. Anyone with a passing interest in the future of wild Atlantic salmon might well wonder what happens when a few of the modified ones, if modification is allowed around here, escape and start to become romantically involved with the remaining wild survivors on the local run.

Three days later, I am on a Zodiac dinghy in yellow wellies and oilskins, heading out to the Loch Duart facility in Badcall Bay, to try to find more answers. It beats a third consecutive day of walking twenty miles in the heat and, anyway, it has taken many months to find a farmer who wanted me anywhere near their pens, let alone on them.

Hazel Wade, newly promoted Operations Director, is keen from the outset to emphasise the importance of their connection to the community as we speed the mile or so across the bay, light playing on the water, and a hundred shades of blue and green stretching out into the distance. 'Just ask how many of the eleven pupils at Scourie Primary School are related to Loch Duart.'[31] It is a forest school, and I had seen the children playing noisily and happily in a wood behind the village earlier in the day.

'Warm water is the main enemy,' she says,[32] as we climb on board the steel gangway that surrounds the pens. 'And before you

---

[31] Highland Council
[32] Warm water and the lower levels of oxygen and higher acidification that stems from it.

ask, we do not yet have a foolproof plan for any invasion of micro jellyfish this year. Nor does anyone. It is something that concerns everyone, and which the industry is working on hard to solve. Micro-nets may be half the answer, but we think that they also compromise water quality as well.' My first impression, looking into the nearest tank, is that it is not as densely crowded as I had expected it to be. The second is that the talk of there being very few 'real' jobs seems wide of the mark; excluding our group, there are four people working on the one rig.[33] The third impression is that I have probably been sent to one of the better operations in the industry. To be fair to Salmon Scotland, who organised the visit for me, it was the farm that I had asked to see, as it involved virtually no detour from my route, and it happened to be the one whose produce was offered to diners at the recent COP26 in Glasgow.[34] And Loch Duart didn't duck my questions about the very high rate of sea lice on their farms in the summer and autumn of 2022.[35]

Hazel talks of predators, mainly net-breaking seals, and how the salmon prefer wind and rain to the bright sunshine that we have today. She says that there is no antibiotic use on the farm, no growth hormones and no sand eels in the feed mix. When I ask her about pre-harvest mortality, she tells me the figures, and then explains that they employ a health manager and two biologists to try to improve things as they go along. After a while, we transfer over to the feed storage tank, from where the pellets are dispersed to the various pens. Down below the water line, she shows me a small room with four black and white CCTV screens being monitored by a

---

[33] Official statistics show a fall from 1,630 full-time jobs (2020) to 1,495 (2021), probably through increased productivity. (Scottish Fish Farm Production Survey 2021. Marine Scotland Science. Scottish Government.)

[34] Since when, I should point out, they have had to agree to drop the word 'sustainable' from their advertising.

[35] In 2022, Loch Duart was one of the worst performing companies in relation to sea lice, with over 35 per cent of its counts breaching best-practice thresholds.

hitherto invisible employee. 'I'm checking that pellets aren't sinking to the bottom of the pen,' the girl says. 'Because if they are, then we are over-feeding, and creating a waste problem down below.' It looked the most boring job I had ever seen, even including the Brussels sprouts I picked for a few weeks as a teenager one winter, but, equally, it displayed an attention to detail that I found striking.

Hazel is sanguine about the opposition her industry faces. 'The only problem I have,' she says, 'is that our critics tend to multiply up our worst-case month on, say, sea lice, and then publicise it as normal. We're not saying that it's perfect, but it would be good if they gave the whole picture. We are initially criticised for not releasing lice numbers; then we get criticised for the numbers themselves. It seems the goalposts are always moving.' She insists that they have had no escapees for years, but then concedes that that's probably not going to interest the critics.

That evening, now in Lochinver, I ask for salmon for the third night running, as I see that it comes from the farm I was on a few hours earlier. The waitress tells me that it is off the menu, so I have the mussels instead, and am so sick that I lose two days of my life, and 6.5 kilos of weight.

Early the following morning, I send a miserable WhatsApp message to my family from the cold tiles of the bathroom floor of my bed and breakfast, and receive a less than sympathetic one back from my eldest son. 'Own goal,' it says with unnecessary honesty. 'There isn't an "R" in the month. Should have stuck to salmon.'

'It's a sign,' I text back, but without knowing of what, and for whom.

So, would I eat farmed salmon again, by choice?

I still find this a complex question to answer and to keep rational. This is not least because this is factory farming, pure and simple, and thus brings with it all the problems associated with a flight from nature, not least the very real threat to livelihoods, nutrition and food security of the coastal communities in West Africa and Latin America, from where the salmon's diet is now extracted in ever-increasing quantities.[36] 'The existence of a market is not proof of a need,' as one commentator put it twenty years ago, 'and permanent damage is not a fair price for short-term gain.'[37] This is an industry that intends to double in size in the next ten years and yet it already has regular pre-harvest mortality rates of around a quarter of its stock, an unthinkable situation in other areas of agriculture. But all the damage that I have learned about – the pesticides, the antibiotics, the lousy welfare, the depletion of other fish stocks to provide food – have to be balanced out by the actual need the eight billion of us have to eat, and for some of that food to be protein; and in salmon we have a nutritious, versatile and relatively low-carbon food that operates, even if inefficiently,[38] on an extraordinarily small footprint of 'land'. And in Badcall Bay, admittedly from a sample of one, I saw for myself the apparent care that was being taken to raise fish, with the environment and their welfare seemingly high on the agenda. Aquaculture is not going to go away, so we need to persuade it to change for the better, something we can all do with our wallets.

Over a coffee on a rainy morning on Loch Crinan a few weeks later, I ask John Aitchison, wildlife filmmaker and respected campaigner on the facts of this issue, what he feels the key problem is, when all the hyperbole is stripped out of the argument. 'It's an adoption of the precautionary principle,' he says. 'That's what I would like to see. But the industry is powerful, and it prefers

---

[36] In Senegal alone, fish consumption declined by 50 per cent between 2009 and 2018, resulting in migration of fishers between West African coastal states. Salmon production accounts for just 3.9 per cent of farmed fish globally, but uses up 58 per cent of all fish oil and 14 per cent of fish meal destined for aquaculture. (Feedbackglobal.org.)

[37] *Sea Change*, Richard Girling.

[38] To get a kilo of salmon to your plate takes around 5 kilos of food if you accept that we generally only eat the fillets.

adaptive management to the precautionary principle;[39] it insists that we mustn't postpone action and growth in default of perfect information.' He adds that there absolutely needs to be a full cumulative impact survey on the existing industry before further damaging growth is allowed. I am struck by his quiet determination to stick to the facts, just as I was struck by Hazel's determination to be a good farmer, and the potential there might be in those two positions for a positive way forward. Wherever I went, I found a widespread view that SEPA is a weak regulator,[40] and a background feeling among some is that this weakness might possibly suit a government for whom the industry is a giant plank of their economic planning for future independence. As it happens, two developments might be about to make the local problem a whole lot more immediate. First, the Norwegian government,[41] who (like Washington State, British Columbia, Argentina, Falkland Islands, Denmark and others) view salmon farming through a rather different prism of risk than the Edinburgh policy makers, are proposing to apply a punitive 25 per cent resource tax on it, which will probably apply more pressure for even more farms within the softer Scottish regime; secondly, the sea is warming year on year, which will only exacerbate salmon health problems in the future. If salmon farming eventually comes to an end, I suspect it will be sea temperatures that take it there, and not regulation or a shrinking market.

---

[39] For balance, when I mentioned the precautionary principle to him, an industry insider replied that if you followed that argument to its logical conclusion, you would never build a house, drive a car or fly in an aeroplane. It is the application of this principle, he said, that is bringing carnage to both Holland's farms and its politics.

[40] Scottish Environment Protection Agency.

[41] The Faroese government is doing the same, but at a rate of 20 per cent, and calculated with the costs of the operations taken into consideration. (*Fish Farmer* magazine, June 2023)

The abstract of a report published in March 2024 tells us that 'Mass mortality events have increased in frequency from 2012 to 2022, particularly in Norway, Canada and the UK', adding that they have also increased in scale, and predicted that the risk profile is hugely affected by 'manufactured risk', a polite way of describing the relentless drive to productivity through the use of technology.[42]

Strangely, the temporary answer may be easier than we think, if only because it is the same answer to most things. Less. If we eat less salmon, much less salmon, expect to pay more for it, raise those fewer fish in smaller, less medicated, better regulated, more enclosed farms, maybe onshore, maybe far out to sea, and feed them on, say, mussels rather than sand eels, Peruvian anchovies and Brazilian soya, we might end up with something that delivers most of the benefits of the industry with fewer of the harms. A moratorium on new farms would help, so that incremental improvements would have to be made to existing facilities, rather than further 'greenfield' expansion.

As with pigs, cattle and chicken, so with fish: if we delude ourselves that factory-farmed protein is genuinely cheap food according to our narrow definition of 'price', and that we have a right to eat it as often as possible, our remaining biodiversity will go on picking up the real bill for the fictional money we are saving. That is the tragedy of the commons, and always has been.

Thus, if you are content to eat low welfare bacon, or broiler chickens whose shabby indoor lives end at just six weeks old, I guess there is precious little logic in your avoiding salmon. The joy of the internet is that you have both the opportunity, and the obligation, to find out if you approve of where it comes from, and not to

---

[42] *Quantitative analysis of mass mortality events in salmon aquaculture shows increasing scale of fish loss events around the world,* Singh et al, *Scientific Reports* journal, March 2024.

be merely seduced by those soft-focus marketing shots of peaceful sea lochs and pile thoughtlessly in.[43]

Buy it, then, if you will, but try your hardest to buy it well. For my own part, I won't: I find that there is simply too much that is too wrong with the very fundamentals of the industry for me to allow myself to support its production.

---

[43] Theoretically, at least. I spent time considering the various certification schemes on offer, with a view to signposting readers towards them. None convinced me enough to do so.

The Birdwatcher

At Ardtoe, by the inlet where the Ardnamurchan peninsula meets Moidart, I meet the worst birdwatcher in the whole world. He is so bad, he makes even a relative amateur like me look like John James Audubon.

He's called Bob. He's from the Borders and he is staying with some cousins, locally. On this drizzly early morning, he has set his stall out on some flat grass just off the little road that leads to the bridge: telescope, tripod, backpack, flask, new bird book. Curiously, he is wearing a bright orange storm jacket and what look suspiciously like loafers on his feet. I ask him cordially if there's 'anything about', which is polite birder code for ensuring that you aren't missing anything good that the other person has worked hard for hours to find. Occasional birders see others of their kind as rivals, not comrades, and find it surprisingly hard to share a good spot. Not Bob. He is delighted with the prospect of company. In fact, if he had a welcome mat in the car, he would run and fetch it and put it down in front of me.

'Not much,' he says, as a curlew flutes cheerfully seaward behind him. 'I'm hoping for eagles and, if I'm really lucky, a capercaillie. I'd love to see one of those. Big stuff is what I want. I'm not so good on the small ones. Can't quite work it all out yet.' He tells me that a capercaillie is the size of a turkey, which is the reason why he wants to see it. In Bob's world, I sense, big is beautiful.

I'm not sure about the frequency of eagles over this bay, but I hardly dare to tell him that his nearest capercaillie is probably around two and a half hours' drive to the north-east, tucked away in some lonely forest in the Cairngorms National Park, unless there's a stuffed one in some local hotel lounge. So instead I tell him that, in half a century of quite committed birdwatching, I have never actually managed to clap eyes on one myself, but that I am always delighted by whatever else is around. He looks a bit crestfallen at the potential loss of his trophy species, so I point out a few shelduck on the estuary, line astern like battleships at a naval review. And then I suggest that he is in a great place just to enjoy the birdlife. A hooded crow flaps past and he asks me excitedly to confirm what it is, and if it is rare, what with that unusual colouring. Rarity is a big deal for him, almost as big as size. To dilute his disappointment at the answer, I also mention that the hoodie has only been recognised as a distinct species since 2002, and that if he'd seen it before then, he couldn't have counted it.

'It's my retirement gift to myself, all this,' he says, hunching his shoulders ever so slightly, as if he might have disappointed me in some way. 'I needed something to get me out of the house and keep my brain functioning. This way, I learn something new every day.

'I've spent all my life with zero interest in the countryside, and zero contribution to nature. Every time I watched one of those nature films going on about decline, I felt bad, like an alcoholic trying to ignore public service health warnings. I suppose this is me trying to do something about it. Maybe I can help my grandchildren get interested, too.'

And suddenly, I get it. He's only doing at 70 what I was lucky enough to do with my dad at seven and, just like me, he will get there. He's not bad, just new. He's not hopeless, but hopeful. And most important, he is swelling by one the ranks of the people who give a damn about our nature, which is good however you look at it.

And if he persuades even one of his grandchildren to actively join nature's pathway, he will have performed a tiny miracle. Nature needs all the human engagement it can get, as it is only the adults that those children will become who can lead us out of the mess that previous generations have created.

And with that, I head slowly westwards into the forests of the remote Ardnamurchan peninsula, musing over just how much Britain needs more Bobs.

THE HONEY POT ISLAND: THE SEA EAGLES OF MULL

3

Late June

> 'The World will never starve for want of wonders, but only from the want of wonder.'
>
> G.K. Chesterton in *Tremendous Trifles*

On the short ferry ride from Kilchoan to Tobermory the following day, I find myself thinking once again about those missing seabirds. A desultory shearwater commutes westwards across insufficient wind for his dynamic soaring, and a couple of guillemots bob on the water as we pass, but that is it.

Many people I speak to on my walk down the west coast have said that they believe it is down to the medium-term effects of bird flu,[1] or over-fishing, both of which have certainly contributed, but I am becoming increasingly convinced that there is something deeper and more permanent at work, something that I will go on to meet time and again on my journey, namely marine heatwaves (MHWs). The science doesn't seem to disagree. Shortly after I returned home from the first of my walks in the summer

---

[1] Also known as Highly Pathogenic Avian Influenza (HPAI).

of 2022, a prominent climate-change service confirmed that the average daily global sea surface temperature had just reached its hottest ever recorded level, of 20.96 degrees centigrade. More worryingly, it had done so at entirely the 'wrong' time of year; March, not August, is when oceans globally are supposed to be at the warmest.[2] And, sure enough, in March 2024, figures provided by the EU Earth Observatory Agency confirmed that the new record had already been broken, and the 21 degree barrier passed. Hotter seas absorb less carbon dioxide and indirectly accelerate sea level rise through thermal expansion, quite apart from promoting harmful algal blooms and the effect they have on marine wildlife.

Water temperatures around Britain are naturally variable from year to year according to the El Niño cycle, largely dependent on ocean currents. In addition, they have recently been warming at the persistent rate of about 1 degree centigrade for every 30 years, but both the pace and the severity has accelerated since the turn of this century. The fact is that the planet now has an energy imbalance, which we might know as climate change, but which the ocean experiences as 'more energy coming in than is going out',[3] and, from that, warming. While this might make for a more comfortable dip at the end of a long day's walk, its long-term effects of desalination, deoxygenation and acidification are causing havoc under the waves which, because we can't see them, tend to pass under the collective radar. As we saw in the last chapter, warming seas make healthy aquaculture much more of a challenge with each passing year, just as warming rivers reduce the spawning success of wild Atlantic salmon, but there is much more to it than that. Coral bleaching and a deterioration in both seagrass meadows and kelp forests takes place, with consequences in reduced

---

[2] EU's Copernicus Climate Change Service. 3 August 2023.
[3] *Blue Machine*, Helen Czerski, Penguin Random House 2023.

carbon burial and sequestration and decreased phytoplankton production.[4] This matters, because phytoplankton is one of the bases of the aquatic food web, and if that decreases and moves northwards, so do its primary (zooplankton and small fish) and secondary (big fish, sharks and whales) consumers. Flying fish and Risso's dolphins appearing regularly off the coast of Cornwall may be good for wildlife-watching business, but they are a worrying signal of what is happening below the surface of the waves. Off the coast, warming water is good news for mackerel, blue fin tuna and the pacific oyster, but bad news for just about everything else, including cod, herring, whiting and sprat. If you happen to be a great skua, Arctic skua or Leach's storm petrel, it is now quite likely you will go locally extinct, possibly fully, sometime in the coming decades.

Technically, MHWs are periods of five days or more when the sea-surface temperature exceeds a local seasonal threshold, and they have doubled in frequency in the last 40 years or so.[5] For now, it seems that it is more about distribution than overall numbers, although seabirds and corals happen to be particularly affected. However, it is the imminent vulnerability of whole ecosystems that worries academics, particularly when populations are already at the warm limit of their range. As an illustration of the problem, some 'plant and animal species have retracted poleward by more than 100 km following severe MHW events'.[6] Hence the conveyor belt of fish heading north.

---

[4] *Ocean Community warming responses explained by thermal affinities and temperature gradients*, Burrows et al. Paper for *Nature Climate Change*, 25 November 2019.

[5] *Marine Heatwaves threaten global biodiversity and the provision of ecosystem services*, Smale et al. Paper for *Nature Climate Change*, 4 March 2019.

[6] In late summer of 2024, scientists were noting a cooling of the Atlantic beyond what might have been expected as La Niña came into control, and were unable to say why.

And where the fish go, so too do the seabirds that follow them. Hence the silent skies.

However, silent skies don't always need to remain silent. Nature sees to that, if only we let her.

Some time back in the early 1980s, a huge and unfamiliar shape started to drift across the breeze from the Isle of Rum and into the airspace above the forests and glens of northern Mull. With a two and a half metre wingspan and the profile of a 'flying barn door',[7] the white-tailed sea eagle was back after an absence of three-quarters of a century. 'Iconic' is a word overused these days to the point of triviality, but in the case of the white-tail it is no exaggeration: their size and sense of presence is out of all proportion to what we are used to seeing in our skies. It is often wrongly assumed that the sea eagles were directly re-introduced to Mull by man; they were in fact travellers from the thriving (and re-introduced) population on Rum in search of new territories.

Sea eagles have hunted and scavenged alongside man since Neolithic times, and they quickly found plenty on Mull to keep them there. Over the next five years or so, various failed attempts were made to breed, and it wasn't until 1985 that a female called Blondie successfully fledged a chick. Having survived a near drowning in his earliest days by paddling himself to the shore overnight when he had over-immersed himself in the loch, the chick and his successors then became the beneficiaries of an island that suddenly understood the enormous potential value of this gift from the north and was determined

---

[7] The delightfully unscientific description that Roy Dennis gave them over 50 years ago when first helping re-introduce them to Rum, and which has stuck.

to protect its eagles. Egg collecting was still a real threat, before the risk of huge fines and possible prison sentences began slowly to restrict it to a small and viciously determined minority, so a species protection officer was appointed and a police operation set up that led inexorably to the Mull Eagle Watch, which continues in the breeding season to this day. This was and is a community that understands the value of its wildlife, specifically its sea eagles, and over the next 30 years the island slowly reinvented itself.

Ecosystems change; they always have. Although it rarely happens, any talk of the sea eagle should also be accompanied by talk of its fellow arrivals, such as goldfinches tempted over by birdfeeders, siskins and crossbills by the new forestry, and of those no longer seen, such as the chough, the black grouse and the grey partridge. Foxes and badgers are long gone, while otters and the long-absent pine martens are back in force, as are feral cats, possibly the most destructive of all residents aside from us. Into this complex tapestry, then, the sea eagle started to weave its precious thread.[8]

As their territories grew in number to twenty and more, so too did the opportunities to attract nature tourists and their cash. When I had been a regular visitor to the island as a boy, there seemed to be a sense of inevitable decline and lack of opportunity about the place that was reflected in a static population of around 2,000 people and relatively few young people choosing to stay to work. Not any more. These days, the population is nearer 3,000 and there are many thriving wildlife and tour businesses operating there, ferrying people to see otters (themselves beneficiaries of the protection and the banning of DDT), puffins, minke whales and much else besides. There has

---

[8] For much more on this, see *Wild Mull: A Natural History of the Island and its People*, Littlewood and Jones.

been a radical change from the nineteenth century, when 'eagle economics' basically meant rich people coming over to shoot them, rather than now, when it is all about the natural history, the fleeting glimpse and the camera.

Above all, the camera.

On a high bluff by the little road from Dervaig to Tobermory, Ewan Miles explains how to tell the difference between a sea eagle and a golden eagle at a distance.[9]

'Look for the flat profile across the top of the wingspan,' he says, 'and that's a sea eagle. The goldie has the shape of a squashed letter M, like a buzzard.'

Ewan originally came up to the island from his native Cumbria as a student in 2010, to help with sealife surveys. Over the years of doing that, he spotted a niche for a personalised guiding service, in which he could make a living by sharing his passion with guests. 'It's a great day at the office,' he says. His first gig was with a family from Liverpool, delighted with their day as it turned out, and 'it sort of went from there'. So much so that, after a couple of years, he found he was working 70 to 80 hours a week, running tours both in the day and evenings, and then coming home to do his paperwork at around midnight. These days he employs two young naturalists and also takes regular trips off the island to other wildlife hotspots like the Cairngorms and Islay. He is one of many, a respected fixture in the economic bounty that nature has inadvertently delivered to the island. His workplace and focus change with the conditions, from the waders hopping between the bladderwrack-covered rocks of Loch Scridain to the hen harriers hunting a beat below the mists of Ben More.

---

[9] Mull also has one of the highest densities of golden eagles in Europe.

As we drive off again in the vague direction the eagle flew, he acknowledges that he has seen many changes even over this relatively short time and is acutely aware of the battle lines that can get drawn, often artificially, between conservationists, farmers, fishermen and others over any number of issues. 'Bracken control for one,' he says. 'Deer management, sheep grazing, disturbance, marine protection areas, creel fishers pointing fingers at the eagles and, above all, the issue of eagles and lambs.' It may be a living, but Ewan sees his team as part of an environmental movement, where the lesson that nature is our life-support system can be drummed gently into each guest. 'The trick is not simply to get as close as you possibly can to something just so that your guest can get the best photo, but rather to allow them to live in the moment whilst they watch something from a respectful distance that is at once routine but utterly amazing.' There is a deep sense in what he says of nature's unexpected vulnerability in the Instagram age, but also of the importance of getting the central message across to people before it is too late. He is convinced that the two are not incompatible.

'Each person we share this with will potentially become a better ambassador for wildlife,' he says. 'The trick is to keep them at a respectful distance.'

The following afternoon, I get a boat to the little island of Lunga, a childhood favourite of mine, to see once again the colonies of breeding seabirds there. Well might there be storm petrels, Manx shearwaters, razorbills, guillemots and shags, but in reality there is only one show in town for the majority of visitors, the charismatic puffins who nest in burrows on a ledge above the makeshift harbour. As a denatured people, we generally want our nature cute, 'charismatic' is the technical term, which is why *fratercula arctica* delights so many. While there is some evidence that the puffins actually time their flights back from the sea to the arrival of the boats (a human is less of a threat than a skua or black-backed gull, whose presence the former tend to discourage), it is shocking to see how close people routinely go to the birds to get the money shot. I watch one lady armed with an enormous camera wriggle her way to within a metre of a bird, blocking it from accessing the chick in its burrow, and another, a middle-aged man, posing with a glass and a little bottle of whisky at the entrance to a burrow that a puffin is patiently trying to access. Unwilling though I am to cause discord, I remonstrate with the latter, as it is so manifestly wrong. Stuart Gibson, who has been a volunteer wildlife guide on these trips for twenty years, is in no doubt of the wider danger of this behaviour.

'When the white-tails came back,' he says, as the two of us watch the puffin watchers watching the puffins, 'they were responsible for putting cars on ferries, heads on hotel pillows and bums on restaurant seats. It has put the island on the map again, and given it a purpose.

'But now I think we are in danger of killing the goose that lays the golden egg. The puffins here can probably cope with it, but when you have people blocking an otter's access up to its holt so that they can get a good face-on, whiskers-in-focus picture, or getting so close to an eagle's nest that both parents take to the air, you really start to fear for the future. Proper wildlife photographers go to the ends of the earth not to cause disturbance, and so should everyone else. At the moment, they are applying their first world rights to nature, and there is a risk that the place ends up like Serengeti, with five groups of tourists looking at the one otter from five different directions.'

'What's the answer?' I ask him, as we make our way carefully down the steep path back to the boat at the back of the party of customers. 'I suppose the horse has bolted.'

'Not at all,' he says, carefully out of earshot of the others. 'Just place a blanket ban on phones and cameras. That would do the trick!' I sense that he is only half joking.

Later, I ask Conor Ryan, a renowned expert on baleen whales and cetaceans who also doubles as a volunteer marine guide, whether the endless parade of whale-watching boats poses any risk to the animals themselves.

'It's the balance of probabilities,' he says. 'More tours going further afield in much bigger, faster and louder boats are more likely to strike whales and, in the long term, persuade them to move away, and effectively be excluded from their own habitat.'

For now, the warming seas are having far more effect on the cetaceans than a few thousand visitors do, as the conveyor belt of the food system heads slowly and inexorably north. Minke whales are undergoing a slow decline in the area, and being replaced more frequently by humpbacks; the two remaining members of the West Coast Community of orca (also known as killer whales) will eventually die out, and be replaced by different ones from the south.

Like Ewan and Stuart, Conor acknowledges the power of the exposure to wildlife in changing behaviours: 'One child, one adult even, seeing an orca for the first time in their lives could change them for ever, and positively.' What worries him is the pace at which tourism is developing, with photographs and a tick-list of species replacing the lived experience of raw nature. Where our forebears were content to watch and wait, for days on end if that was what it took, social media has led to our modern dopamine neurotransmitters needing all of it, and right now. 'They used to come here for one week, maybe two. But now it tends to be two or three days and then head off to the next highlight somewhere else.' In these fast-paced, adrenalin-fuelled days of instant gratification, even a place with the treasures of Mull has to fight for a few days' attention, and maybe that factor is really what is behind the tensions. These days, nature is an uneasy branch of the entertainment industry.

The nature of Mull has created between 100 and 160 full-time jobs on an island that badly needs them, and it brings in about £8 million of local income annually.[10] It is a genie that is not going to be shoved back into the bottle, and nor should it, but the time may be approaching when a more enforced code of practice has to be put in place.

For a few days, I just tramp the island coast on my own, enjoying the unspoiled beauty of the place.

Like all islands, the Mull below my feet is an amalgam of four forces – geology, climate, ocean and mankind – that have gradually shaped it. Possessing both some of the oldest and yet youngest rocks on earth, often within a few miles of each other, you can be standing on 2.5 billion years of history on Iona, and then something 95 per cent newer just outside Bunessan. Never better than Class 5 farmland ('capable of use as improved grassland'), it has generally been given over to livestock, into the gaps between which have ebbed and flowed the elements of wildlife that make it what it now is.

The Mull of my childhood, the Mull of yesterday's newspapers and tomorrow's mackerel, of annoying walks and even more annoying adults expecting a boy to be interested in cutting bracken or pulling ragwort, has transformed into a place of almost unimaginable treasures. These pedestrianised treasures of mine must be shared with what seems to be every white camper van on earth and with coaches hurtling their way down the single track Glen Road towards Iona with their cargo of temporary pilgrims. Long before my grandmother came, that last stretch towards Iona was one of western Scotland's principal coffin roads, where the recently departed would be carried across the Ross of Mull by

---

[10] RSPB Study. *White-tailed eagles bring tourism boost to Mull*, March 2022

a different sort of pilgrim, towards a resting place beyond the Abbey's Street of the Dead and into the earth of the machair to the west beyond. Always to the western shore, so they say, to be nearer the setting sun. 'Rubbish,' say the cynics, 'it's just that the earth was softer there.'[11]

New life is everywhere on this summer day. In the intertidal shallows of Loch Scridain, I watch an otter playing with her young among the bladderwrack seaweed and orange-flecked basalt rocks under the flight of a watchful golden eagle and the trill of a concerned oystercatcher. Over on the other side at Ardcrishnish is an embryonic community-owned seaweed-growing business, part of a vibrant island-wide effort to put the community not just at the heart of living here, but actually starting to influence what happens here economically as well. In Bunessan, I sit on the sea wall and stare at a grey heron painstakingly gulping down, bit by wriggling bit, an unfeasibly huge eel that refuses to die at all, let alone quickly. Over a coffee at Fionnphort, it's a cock blackbird almost between my feet, common as muck, maybe, but bold as brass and magnificently dark.

At that moment, 40 pilgrim tourists decant noisily from their tour bus above the slipway to the Iona ferry and are shepherded quickly towards the waiting boat by their guide. 'There are corncrakes over there if you know where to look for them,' I say excitedly to two of them, who stop to chat. For a second, the older of the two looks at me in complete confusion, as if I had offered to show him some pornography, and then he smiles weakly. It seems corncrakes weren't in the brochure.

For this is about Columbus, not corncrakes, abbeys not animals, and they pass quickly by. Two hours is all Iona's 1,500-year history manages to prise out for itself from their busy schedule. They are

---

[11] For more on this oddly fascinating topic, read Ian Bradley's *The Coffin Roads* (Birlinn, 2022).

on a deadline and need to be back in their Oban hotel by nightfall. If they are in luck, they will be piped off the boat when they get there.

That evening, at Macgochans Bar in Tobermory, I have arranged to meet Moray Finch, the general manager of the Mull and Iona Community Trust. I want to learn more about how the coastal community best lives around the honeypot of tourism, and he should know. He was brought in to facilitate exactly that.

'Community-led ranger service, charity shops, business development, a marine pontoon, a community forest ...' he ticks off the initiatives that he is working on. 'But clearly affordable housing is the key one.' Pick any coastal settlement in Britain, and that will probably be the case.

A glance through the ill-lit estate agent window backs up his assertion that house prices on Mull are high, and are continuing to rise. Spiking after Covid when people from elsewhere mysteriously realised that they didn't fancy being restricted to living only in cities, the current average house price in Mull is only just below the UK average, and much higher than it was at the turn of the millennium. Add that to an absentee house ownership rate of around 35 per cent, and it is scarcely surprising that the very people who are needed to fuel the economy currently struggle to find somewhere even to lay their head. While he understands why Muileachs may resent economic changes that have failed to bring equivalent changes to incomes, Finch refuses to demonise the second-home owners, whom he says create jobs for tradespeople and may well also be local themselves. Ironically, one of the biggest problems is the lack of changeover staff for the holiday cottages, where that particular job rate has had to double in recent times in order to keep the service going at all. Instead, he is concentrating his efforts on social interventions, such as building four low-cost houses at Ulva and working actively to keep a local primary school going.

'Conservation is not enough,' he says. 'We need to do restoration; we need to have the same people doing different things.' He is justifiably proud of his two community-led rangers, whose role it is to go around the hotspots and do exactly what Stuart and Ewan would like to see them do, helping nature watchers understand how to coexist with the wildlife they are watching, and stop them going too far.

Probably the biggest issue facing him on behalf of the islanders at the present time is the dire state of the ferry service, for which Caledonian MacBrayne famously has the monopoly. For people who have not spent time around the Western Isles, it is hard to imagine to what extent normality depends on the smooth working of the ferry network that connects the 23 islands. Reliable ferries are the lifeblood of island communities. Those familiar, red-funnelled ships are the daily inhalation and exhalation of people and goods that keep the show on the road, the artery that keeps them connected to the rest of Scotland. It has been anything but reliable in recent years, and many boats are simply not fit for purpose.[12] For the visiting tourist it is an annoying enough inconvenience, but for, say, a local livestock haulier or a cancer patient with an appointment to keep on the mainland, it is far more serious. Day after day, boats are delayed, changed or cancelled at the drop of a hat, and at the time of writing there seems little prospect of them improving any time soon. The operating company uses boats owned indirectly by the government, and it is hard to see some entrepreneur being able to overturn both the structure and the monopoly. Finch has organised feasibility studies for an independent operation on the Oban–Craignure service, but because it is 'bundled' in with services to the other 21 island

---

[12] 'Pitch control issues', whatever they might be, caused the cancellation of the Lochboisdale (South Uist) to Oban ferry in June 2023 for an entire month when I was on the island, with long and tortuous alternatives the only option. To say locals were inconvenienced is a reckless understatement.

destinations, he doesn't believe that it will happen. This concerns him, as he is a strong believer in de-centralisation.

I ask him if the island could take further growth, an even greater population, because the impetus for it is everywhere you look. He answers obliquely by referring to a scheme for a development of 90 new houses for Craignure. Over twice the size of the Isle of Wight, which has 50 times the number of people, there is clearly the space for more on Mull, but only if it is accompanied by sustainable year-round jobs and the right infrastructure.

Talking to locals who may be less diplomatic than Moray Finch, it would help if more of the thousands of people arriving in camper vans at least shopped on the island, rather than the mainland, and spread the wealth that way. With careful management, the honeypot can probably grow. There are issues, of course there are, but you get the feeling that what is really needed is a lot more decision making delegated to the islanders.

And a ferry that works.

On my last full day on the island, I go to sea again to seek out more white-tails, this time in the company of the man who has championed them for just about every one of the 38 years they have been here. After all, they are why I am on the island. And he, to a large extent, is why they are here, too.

Dave Sexton came to Mull in 1985, as the RSPB's newly appointed species protection officer for the white-tailed eagle, and he kept coming back, eventually for good. If anyone can claim to represent the bird on this island, he can. For the last eighteen months, he has been monitoring the effects of avian flu on his charges, most particularly the marked reduction in fledging success since the disease first made its presence felt in the autumn wildfowl migrations of 2021. He is sanguine but concerned about possible future mutations that may worsen the situation. We have talked eagles together many times before, but this time

I principally want to ask him about the elephant in the eyrie, the reputation of the bird as a voracious killer of lambs, for it is this that is the seemingly unbridgeable divide between farming and conservation. I mention a sheep farmer on the west of the island that we both know, who says that she has seen with her own eyes the taking of healthy lambs from her flock. I have met many others with the same story on my journey south from Cape Wrath.

'No one denies that can happen,' he says, 'or that there are real human costs, but there is also a basic ecological misunderstanding and unforgivable misrepresentation about how the eagles hunt and feed. The vast majority have a very varied natural diet of seabirds, fish and mammals, and they do not seek out lambs. They will sit for five or six hours looking for opportunities, often watching hooded crows, great black-backed gulls and ravens who themselves have made a living off lambing since the dawn of farming. That's what brings them in, especially the young birds. They have scavenged off mankind since Neolithic times, and now all of a sudden, just when we have brought them back from the brink of extinction, back to areas they have historically always been present, certain people and organisations want rid of them again because they are blamed for disrupting our sheep farming.' He adds that the reputation the bird has among farmers is, as far as he can see, a significant miscarriage of wildlife justice.

He continues: 'The eagles are wrongly accused of countless lamb deaths based on spurious evidence: things like seeing eagles carrying lambs, being seen feeding on dead lambs, farmers finding carcasses which have been fed on, or "black loss" [when lambs just vanish into thin air]. It is as if organisations like NFU Scotland [National Farmers Union] completely disregard the basic sea eagle ecology of scavenging. The eagles use a survival technique known as klepto-parasitism. They steal fish off otters, make gannets throw up their mackerel and routinely steal the remains of lambs from other predators. And that's when they get photographed.'

Even though the dominant element in the eagle's diet at lambing time seems to be breeding seabirds, hares and goslings, this is not a problem that is conveniently going to go away, especially with an increasing population of eagles and a decreasing population of seabirds. Recent work that has looked at remains left in nests, most of which do not contain lambs at all, has reported a trend that indicates that lambs are a decreasingly important food for white-tailed eagles,[13] and that these losses can have a measurable long-term impact on the viability of sheep farms. This is not a problem elsewhere in Europe, possibly because the sea-eagle breeding cycle in Scotland coincides so directly with peak West Coast lambing periods.

Some possible solutions are being trialled elsewhere, such as diversionary feeding and Italian Maremma sheepdogs that could be embedded in the flock and trained to chase off eagles, but it is fundamentally a problem that rumbles on. As you would expect, there are endless working groups, management schemes and even the inevitable national stakeholder group, but the truth that dare not speak its name is the one that rears its head at every turn, the brutal primacy of our own species' needs over every other one. This human exceptionalism will be an abiding trope of my journey for the next year. In a country that has lost around 50 per cent of its biodiversity in the last 70 years,[14] and is ranked 189 out of 240 for biodiversity intactness,[15] there is little sign as yet that we are prepared to take the difficult decisions to adapt our own lives to nature, rather than insisting nature continues to do it for us. Like paying reality prices for our meat, eating less of it and, from that, decreasing our livestock numbers.[16] When it comes down to it, our needs have to win, every time.

---

[13] *The Breeding Season Diet of White-Tailed Eagles in Scotland*, Reid et al. Paper for *Scottish Birds*, 4 Dec 2023.

[14] As opposed, for example, to Canada and Finland, which have 89.3 and 88.6 per cent of their biodiversity left intact respectively. Natural History Museum paper, 2020.

[15] Natural History Museum, 2022.

[16] A policy that the influential NFU explicitly ruled out in its submission to 2018 Committee on Climate Change's call for evidence.

Out of the corner of my eye as we head slowly back over the wine-dark sea towards Tobermory, I spot a dark shape slanting towards us from the west. Dave and the rest of our small party are looking the other way at some activity over the hills at Calgary, which means that, for a fraction of a second, she is mine alone. For an instant, I find I can't speak. Her size seems out of all proportion to the local birdlife, too huge for binoculars to be of any use, too personal for a photograph, and she is heading straight over our boat, darkening the airspace maybe 30 feet above us. I am generally wary of exaggerating the emotion of nature, but this time my breath catches in my throat with the awe of the moment, and I find myself gasping. I have been seeing these birds for weeks now, but generally only far away, and high up; this one so close to me is like a glimpse over the wall and into the secret garden. I tell the others and, for a few seconds, we watch her pass right over us. We see every movement of her huge head, her yellow eyes scanning us from above, and pick out the details of her vast, flesh-tearing bill. We see the great plank of her two and a half metre wings, so huge that they are almost an obstacle to light, and the end feathers twitching in the breeze of her flight. We see her huge, slow wingbeats, 'like someone shaking a doormat'.[17] As she heads away from us, the white tail fans out behind her, and only then does anyone say anything. It was as if she had, for a brief moment, awakened in us some ancestral memory of how our skies once were and now are again.

In our Anthropocene world, her presence here is nothing short of a miracle, and a sign of the good that we can do, if we have a mind to. As a species ourselves, we have some tough decisions to take if we are going to keep her here.

---

[17] *Wild Air*, James Macdonald Lockhart, 4th Estate, 2023.

Gulls

Years ago, I was asked to read a passage from Richard Bach's *Jonathan Livingston Seagull* at the memorial service for a young friend who had died an early and violent death far from home.

The story was to be a reflection, her brave parents had insisted, of her sense of adventure, of freedom and her determination to rise above what was expected of a girl like her back in those days.

Inevitably, as I now progress south, gulls become a motif for my own journey, just as their calls have become part of its soundtrack. I will see them just about every day of the coming year. Of the six types that I see and hear regularly,[18] it is the crazy, playground scream of the kittiwakes on the cliffs that delight me the most and wake me from any reverie that I might have fallen into.

The best angle to see gulls is from above, which is probably why I walk along cliff tops whenever I can. From that perspective, you can appreciate every twitch of the wing and tail as the bird compensates for the updrafts and downdrafts of this liminal habitat, and see the tiny movements of its head as it searches out opportunities. And somehow, looking down is so much more comfortable than looking up.

One day, on some low crags just south of Oban, I sit and watch a herring gull playing in the breeze. For me to say 'playing' is to ascribe that gull with a motivation that is not only highly unlikely, but also one I can't begin to understand. All I know is that it isn't obviously carrying out flight for one of the classic purposes of feeding, breeding or avoiding predators, and that it therefore fits conveniently into my anthropogenic vision of what looks like fun,

---

[18] Greater and lesser black-backed gull; herring gull; common gull; black-headed gull; and kittiwake, to which you could add the occasional Mediterranean gull and glaucous gull.

which is good enough for me. Not everything in nature is explicable by science alone.

Then another feeling comes, seemingly from nowhere. Thirty years after that memorial service, the playing of that gull in the wind reconnects me with that long-gone friend, and I find myself understanding for the first time precisely why I was asked to read from that book in the first place.

For the rest of the year, I am more tuned to gulls than ever before.

RESTORING THE COMMONS: FIRST AID FOR SEABEDS IN ARGYLL AND BEYOND 4

Early July

'Scallop dredging is like cutting down a rainforest to catch a parrot.'

Calum Roberts, Marine Biologist

The world unfolds for me on footpaths, drove roads, beaches and tracks at walking pace, and I notice things that I would never notice on a bike or in a car. Having spotted them, I can then think about them at leisure, often without consciously trying: an apple tree by a passing place that maybe arose from a long-discarded core, a hen harrier drafting his silvery way across a bit of heath, a flag iris where it shouldn't be, on top of a drystone wall.

Freed from the tyranny of finding a permitted place to drive and park my car in a crowded world, walking provides the subversive opposite, in that I go more or less where I want. In a car, evolution just brings me backache, frustration and a dose of other people's anger, whereas walking is the easiest and cheapest form of action. At exactly the same time as being a channel for activism, walking also manages to be an opportunity to feel good achieving nothing. 'Thinking is generally thought of as doing nothing in a production-oriented culture,' suggests one of modern America's most committed walkers, 'and doing nothing is hard to do. It's best done by disguising it as doing something, and the something closest to doing nothing is walking.'[1]

For most of modern history, walking in public was something only truly available to half the population, my half, which is an unfortunate and shameful privilege. For the other half, women, it happens to have been an activity largely denied to them by a mix of tradition, rules and the needs of personal safety. It is only recently that it has been edging towards more inclusivity, just at the point that walking itself is being slowly replaced by a combination of the wheel, for those who have to move themselves some distance, and the screen, for those who don't. Evidence suggests that our pre-agrarian forebears had evolved to walk somewhere between ten and fifteen miles a day, a figure that has now declined to around two and a half,[2] meaning that many, even most of us, aren't even walking that. The radius of our acceptable walking distance is shrinking all the time while it curiously becomes something we pay to do in an airless gym, rather than for free in the city streets. For nature, this clearly matters, in that the less we see it, the less we know it – and the less we know it, the less we eventually care about it. Who mourns the disappearing curlew who has never even seen or heard one through the closed car window? *Solvitur ambulando*, as the Romans used to say. You can work it out by walking.[3]

But even with walking, you can only see what is there in plain sight. And most of what lies just over my right shoulder on this

---

[1] *Wanderlust: A History of Walking*, Rebecca Solnit, Granta, 2022.

[2] 5,117 steps daily, according to a paper in *Medicine and Science in Sports*, 'Pedometer-measured Physical Activity and Health Behaviors in US Adults'. Oct 2010.

[3] Quoted from *Nature Cure* by Richard Mabey.

long journey around the coast is below the waves, and out of view, because nearly three-quarters of the earth's solid surface is seabed.

On average, the seabed lies over two miles below the surface of the sea,[4] which means that most of it is beyond even our human ability to mess up, or at least it probably was until the advent of deep-sea mining. The part that is most vulnerable to human activity is the continental shelf, an area that in Britain's case extends around a hundred miles out into the Atlantic Ocean and includes the entire North Sea. Depending on exactly where you are, this has an average depth of between 90 and 150 metres, and it surprises many that the first twelve nautical miles of it belong in perpetuity to the Crown. From mineral rights to wind turbine royalties, and carbon storage licences to undersea pipelines, this is a giant self-licking financial lollipop that contributes generously to an annual profit for the Crown Estates of £442 million,[5] of which the monarch gets to keep 25 per cent.[6] It is an ownership structure that generally fails to delight conservationists and republicans alike, but which may actually be the closest we will ever get to having a coherent policy for our seabed.

If the Crown and the rest of us manage to be good stewards of that vast underwater area, we are amply rewarded by the many useful contributions it can make. This includes the giant kelp forests and sea grass meadows that protect our coast from storm surges, provide nurseries for fish and capture and store carbon

---

[4] Average depth of 3,688 metres, according to the National Oceanic and Atmospheric Administration.

[5] 2022/23 figure. Over the last decade, this has amounted to £3.2 billion. As the income from offshore turbine licences soar, so the government will reduce the percentage that the monarch keeps, even though the figure itself will rise inexorably. Sadly, no one from the Crown Estates found the time to explain their future plans to me.

[6] Other than in Scotland, where it all goes to the public purse.

dioxide; the sediment layer itself provides the same carbon storage function, and also provides a home for millions of small fish and invertebrates that feed the fish that we ourselves catch and eat. Not least, we site wind turbines on it and extract around 200 million tons of primary aggregate (sand, gravel and crushed rock) from it annually, for use in roads, buildings and sea defences.[7] We might not be able to see it, but it is a vital resource which rewards being well looked after.

Only it has not been well looked after. Much of it is in a truly shocking state that we usually only become aware of during Sunday evening nature programmes on our televisions. The dragging of weighted fishing equipment across the seabed (bottom-trawling or dredging) is widespread, including in supposedly protected areas,[8] shattering the structure and life on the sea floor. Ninety per cent of seagrass has vanished and most of Britain's 26,000 square miles of kelp forests are likely to be gone by 2100.[9] The list of current stresses on our seabed and its marine biodiversity is daunting: ocean warming, sea level rise, plastic pollution, alien species, overexploitation, habitat destruction and acidification, to name but a few. It is not surprising that it has become an important front in the battle to save our nature.

I walk on south, always in search of the people who are trying to do exactly that. And as I go, I can make out, through the sad detritus of our human damage, a thousand points of light gleaming back at me through the gloom.

---

[7] The Role of imports to UK aggregates supply. British Geological Survey report for Office of Deputy Prime Minister, 2005.

[8] 'Fishing Industry still "bulldozing" seabed in 90% of UK marine protected areas', *Guardian* article, 31 May 2022.

[9] 'Queen Elizabeth owns most of the UK seabed. That's slowing conservation work', *National Geographic* article, 7 June 2022.

Twenty miles south of Oban, close to the comforting, clanking halyards of the Ardfern Yacht Centre, I seek out and find one of those points of light.

In conservation, lots of things are complicated. This isn't. A small group of people from the local community are, bit by bit, simply mending something that someone else has broken. The seabed of Loch Craignish, like so many inshore waters around Britain, has slowly been degraded by disease and a raft of human activities, including scallop dredging and anchor dragging, over the last 50 years, until the life within it has largely ebbed away. This deterioration was an almost direct consequence of the decision to remove the hundred-year-old three-mile ban on trawling and dredging in 1984, in reaction to a general shortage of fish and therefore a struggling industry.[10] Thirty years later, the effects, although invisible under the water, were grim. Where once you could have caught a cod in the clear water within yards of the shore, by 2010 you couldn't even peer far enough through the murk to see the emptiness below, let alone find a cod; where the rocks in the bay had once been white with gulls, you'd have been lucky to find so much as a single oystercatcher. From the scouring rake marks on the seabed to the vacant skies and the collapsed salmon runs, this pristine treasure had been brutally degraded. While the government publicly wanted to do something about it, all decisions were made centrally and generally only in the interests of business, and then, when the Marine Protected Area (MPA) finally arrived in the area, it didn't even include the loch.

'It started with a simple feeling that we deserved better,' says Philip Price as we potter out into the morning loch under the power of the disconcertingly quiet electric outboard. By the age of 40, he'd already been a builder, a tree surgeon, a photographer and a wildlife

---

[10] A story well told in the film *The Limit*, produced by campaigning organisation Our Seas Need Your Support, www.ourseas.scot

guide, and he not only knew the area well, but also felt a responsibility to articulate the community's sadness about their loch with more than words. 'We set up a petition, but nothing happened. Then, in about 2015, we decided to chuck a few native oysters in to see what would happen.' He explained that oysters create reefs that can act as fish nurseries, and that they feed on algae that they capture by straining and cleaning as much as 50 gallons of water a day, which would obviously help the water quality. 'Then I met Danny Renton, a local documentary maker, at an oyster conference; he had connections with funding streams and we gradually got some money in. First from the National Lottery Heritage Fund, and then from other sources. I guess things just went from there. These days we have grown from one part-timer to six full-time staff, with a satellite project up near Ullapool.'

The Seawilding project was up and running. With half a million oysters on the seabed, there's a lot of filtering going on, and there's the potential of a small income stream from local restaurants and fishmongers out in the future.[11] He opens one of the floating nurseries and fishes out a single oyster, laying it before me in the flat of his hand. 'Look at this!' he says. 'Look how many marine life forms are on just this one shell.' Once I look at it, I can see what he means – there are miniature worms all over just that one shell. In ecosystems, everything is eventually connected to everything else, so it is by no means just the oyster population that is increasing.

'The oysters led on naturally to seagrass,' he continues, before we break off to watch an arctic tern's delicate dance between the sky above and water below. He explains the significant ecosystem services that a healthy seagrass meadow provides, from sequestering and storing carbon to protection from storm surges and providing a nursery for fish to breed in.

'Why the link?' I ask. 'You know, between seagrass and oysters?'

---

[11] For more, and how to get involved, visit www.seawilding.org.

'Seagrass, amongst many other attributes, reduces the acidity of water, which helps the oysters; and oysters, by naturally clearing the water, enable and improve photosynthesis for seagrass. It's the ultimate symbiotic relationship.' These days, as they head towards a million oysters sometime in the next couple of years, they have also planted half a hectare of seagrass, and are planning up to four.'

He senses that I don't think half a hectare is all that much. 'All of this stuff only works in the long term if it is scaled up around hundreds of coastal locations all round Britain. Right now, that's starting to happen, even if slowly.' He lists various current projects in places as far apart as Bangor in North Wales to the Isle of Wight, from Bognor to the Firth of Forth. 'There's as much as 80 hectares of suitable habitat just in this loch,' he adds. He is right. In the world of marine conservation, seagrass is a decidedly sexy area these days.

Initiatives like Seawilding are on the front end of the science, too. They have to be. Working with the academics from Project Seagrass in South Wales, they are beginning to trial different ways of planting the meadows to get a faster and better result. Right now, Philip explains, the current method, placing seeds on the seabed in small, weighted hessian bags, while effective, is quite labour intensive and relatively slow to take effect, so they are trialling new seed germination ideas and various planting methodologies, from direct seed injection to sod transplants and, now, even rhizome transplanting, where whole plants are shifted from donor meadows. As with all farming, the process is reliant on the season, so for large parts of the year the site is relatively quiet, at least on the face of it.

As we pick our way back through the yachts at anchor, I ask him, as I ask just about every conservationist I talk to, what the obstacles are to getting the job done.

'Surprisingly, one of the biggest barriers is legislation. It's very frustrating. Do you realise that we even need to get a construction licence to place hessian bags of seagrass seeds on the seabed? I mean, for replanting what was always here until we destroyed it. Then you can add in all the destructive practices that still go on just over the horizon, or not even that far: dredging, bottom trawling, salmon farming. They deny it's destructive but ...' He lets the sentence drop, leaving me to work out who 'they' are and what the 'but' is.

'What about funding?'

'That's almost the easy bit,' he smiles. He's right. Wherever I have gone, funding is normally a lesser problem than you might think in these challenging financial times. With pressure on companies to burnish their ESG credentials,[12] and a sharp shift towards conservation and biodiversity net gain (BNG) among grant-givers, the money is generally available to those who can prove benefit and, as the Seawilding team has successfully done, grab the PR opportunities as they fly by.

'And people?'

'On top of our six paid staff, we have a small army of volunteers, both from the community and further afield. For me it's one of the best features of the job, as we get the benefit of their labour, and the world then gets the benefit of their new knowledge and their activism.'

'But I'll tell you what I'd also like to see,' he adds as he ties off the little boat to the boardwalk. 'A daily charge, say £3 a night, for the local tourist infrastructure. Just think of how many visitors enjoy this beauty.' he waves his arm across the loch, 'what they can learn from it, and what conservation we could collectively do with that money'. I tell him that I think it's a great idea, even if all it manages to do is start to address the glaring asymmetry between visitor and visited.

But for me, it is in Philip's hurrying off from the marina on his bike that the real key to this and many other projects reveals itself.

---

[12] Environmental, social and corporate governance, a set of criteria beyond mere profit under which investors and others look at companies. When not used for naked greenwashing, it can be a useful bridge between companies and the wider community, building trust and engagement in the process.

He has a 2pm appointment at the local primary school, to talk to and enthuse the children there about Seawilding, how they can be involved and how it relates to their community. Someone once called this kind of activity 'outreach', which has always seemed to me a sterile, invented word more suited to politics, and I wish I could think of a better one. Meanwhile, Seawilding works because it is providing real change and is the honest expression of one community's determination to make decisions for themselves. Philip, Danny and all the other staff and volunteers who have been involved no longer have to sit and hope that someone else does something.

Out there, below the flooding tide of that beautiful inlet, is practical activism at work. And, far more than words, it is activism that will determine the future of our coastline.

A few days later, I learn again that often just doing nothing at all is the best approach for the seabed, because sometimes all nature needs is for us to take our feet off its throat.

The sun is back out, and I am now in the little village of Lamlash on the Isle of Arran, southernmost of the Hebridean archipelago and the ninth island of my journey. While I'm doing what I seem to spend quite a lot of my time doing these days, namely staring out to sea at emptiness, I am also looking at a tiny but vital piece of the history of humanity starting to mend the seabed that it has been hitherto busy breaking.

Sophie Plant, who works on communication for the local conservation organisation COAST, is explaining the background of that invisible miracle to me. In the last years before the millennium, she says, two local scallop divers, Howard Wood and Don MacNeish, were becoming more and more concerned about what they were seeing, or more precisely what they were *not* seeing, in the waters between the village and the nearby Buddhist retreat of Holy Island. Just like at Craignish, years of overfishing and bottom trawling, prompted by the

1984 removal of the ban on bottom trawling within three miles of the coast, had left the seabed raked and bare, and the once abundant sea life almost entirely absent. You didn't even need to go under water to establish this: nothing demonstrated it more starkly than the shrinking catch sizes themselves, whether of the trawlers or the hobby anglers; in twenty short years, 40 recreational angling boats in the Clyde had become just the one. The long-running Lamlash fishing festival which, in 1974, had landed over 16,000 pounds of fish, produced just 200 pounds in 1994 and was never run again.

'As part of their campaign,' says Sophie, 'the two men set up the Community of Arran Seabed Trust (COAST)[13] and campaigned for the next thirteen years for the establishment of a no-take zone in which marine life could gradually recover.'

We look out into the morning sun at the achingly beautiful bay, and to Holy Island beyond. Whether by chance or through what now lies below, the white flecks of seabirds punctuate the view wherever we look. It makes a beguiling and rather appropriate backdrop to the regenerative conversation we are having.

Sophie tells me about the battle to get it going. As all these things tend to be, indeed as most success stories in British conservation tend to be, it was initially a long, hard slog in the face of apathy, inertia and vested interests, starting with the basic principle of just persuading the Scottish government that they had a duty to manage the seas sustainably and in the public interest. It helped that they had already found important habitats under the sea like maerl (a pink coralline seaweed) and large meadows of seagrass. 'Getting locals on side through public meetings and education was key, as was using underwater photography to show people exactly what had happened to their marine neighbourhood, and how bad things were,' she says.

A small site was chosen in conjunction with the local registered fishermen that was logical, easily defined and easily policed, and

---

[13] For more on the organisation, see www.arrancoast.com

then endless letters were exchanged, and meetings held with MSPs and the Edinburgh government, all the time under the gradual growth in global understanding that things simply couldn't go on as they were. It took thirteen years of rejection, persuasion and lobbying until, on 19 September 2008, they finally achieved their goal. The Lamlash Bay No Take Zone, where no fish or shellfish can be taken from its waters including the shore area, became the first of its kind in Scotland, and the first community-based one anywhere in Britain. It is hard to overemphasise the achievement of the David vs Goliath campaign the team ran, and how influential it has gone on to be with others.

At 2.7 square kilometres, the zone itself is a tiny thing in the great scheme of the ocean, but the ecological principle that underpins its success is a strong one, which is that its borders are leaky, that the young fish that hatch within it must quickly migrate outwards or drift out there in the current, and the ones that are already mature grow in size and eventually move out themselves, too. The No Take Zone (NTZ) itself is more like a generator of life than some notional part of a wider sea, and fifteen years on the results have been little less than extraordinary. Most importantly, it shows exactly what could be achieved if this kind of exercise was scaled up and replicated thoughtfully elsewhere, not least for the increased potential for the fishing industry. Researchers from York University have discovered seabed biodiversity increasing by up to 50 per cent and that the size, fertility and abundance of commercial species such as lobsters and scallops is significantly better within the NTZ than outside it. Lobsters are also larger in the reserve than outside it, heavier, more fertile (27 per cent more eggs per lobster than outside) and longer lived.[14] This is no surprise, as it is exactly what

---

[14] The Influence of the Lamlash Bay no-take zone, Firth of Clyde, on spatial and temporal variation in the recovery of commercially exploited *crustaceans*. Dissertation by Eilis Crimmins. University of York.

has also happened down at Lundy Island in the Bristol Channel, which was the UK's first ever NTZ. Cod nurseries have reappeared, the density of king scallops has risen by nearly four times from 6 to 23 per 100 metres squared, and lobsters have thrived to the extent that they are now migrating out into the wider Clyde Estuary.

All these things are, at their best, a balancing act. Making a living from the sea has become a precarious business in the last 50 years, and there are quite as many good fishermen trying to earn their crust sustainably as there are wreckers who'll screw up the commons and then move on to the next one. This has made marine conservation a delicate local issue, where the same community that is wanting to change things, generally meaning a reduction in extraction, is also wanting to earn its living from the bounty of the sea, generally meaning continuing to harvest. So while businesses have a responsibility to stop behaving unsustainably, conservationists need to realise that, for every action, there is an equal and opposite reaction, and that someone is probably being asked to reduce their earnings to accommodate the change. Equally, the huge lobsters wandering their way across the seabed and out of the NTZ are fruits of the investment the local fishermen made, and living proof that sustainable management can work for everyone. It's a wonder they don't do this everywhere.

Or not. In 2022 the Scottish government launched a new public consultation to gather views on Highly Protected Marine Areas (HPMAs) with the aim of designating 10 per cent of Scotland's seas as HPMAs by 2026. In the summer of 2023, they gave the whole HPMA can a healthy kick down the road by delaying it indefinitely, caught, as they clearly were, between a commitment they had made to their Green Party partners when they entered coalition with them, and the alarm and anger of fishing communities at the prospect of losing valuable income through the application of arbitrary area bans in their core fishing grounds. An HPMA that prevents an entire island community from fishing locally is not unlike keeping a class of schoolchildren in detention because one of them is talking.

It's complicated, as I will learn again further south, so I find myself wondering what would happen if the government started by simply banning all scallop dredging and bottom trawling everywhere, and then seeing what effect that had before painting numerous extra arbitrary lines on the surface of the sea. Even this thought turns out to be controversial: new research from the University of Essex shows that one set of man-made changes to the seabed (silting) might actually benefit from another (dredging) in order to harrow the nutrients below.

Meanwhile, COAST has gone from strength to strength. They employ three people full-time and one part-time, and have two full-time seasonal positions, including the skipper of their new boat, RV *Coast Explorer*, which allows them to engage more deeply in citizen science projects, marine research and a whole range of other activities. And, as with their colleagues at Seawilding up at Ardfern, they have the community at the heart of everything that they do, not least in their relevance to a generation of children growing up alongside them.

Maybe, I found myself thinking on the ferry to Ardrossan on the mainland, it is in clear and sustained community actions that the future health largely relies. Get that right, and even governments might sit up and take notice.

The Plastic Collector

Ravenglass; Cumbria

At first, all that catches my eye is a fishing box full of upright cigarette lighters.

I have been walking south from Whitehaven discovering, as all coast walkers eventually do, that the King Charles III Coast Path is rather more of a complete coastal entity on paper than it is on the ground, at least it is on its Cumbrian section. Having therefore had to cut a mile or so inland at various annoying points, I have reached the little village of Ravenglass much later and rather more exasperated than planned. There was a train to catch, but I have missed it.

The box of lighters is in a front garden of the neat street that leads to the estuary. Then I do a double take. Alongside it is an orange bucket of Crocs, a wooden box of ear buds and a washing basket of shotgun cartridges. Above it sit a couple of boxes of branded footwear. Then I realise that the entire garden is covered in recovered plastic, sorted, grouped and then displayed neatly around the hull of an old fishing boat. Behind a hedge there is more: balls, toys, applicators, greasing tubes, asthma inhalers, fishing nets and more, all grouped and displayed as a giant installation. Indeed, while there is no more than the odd square foot of his garden not covered in coloured plastic, there is an understated elegance to it, which gives me a feeling that if Ai Weiwei had created it, it wouldn't be out of place in the Tate Modern. There is a note attached to the boat explaining that this had all been collected from local beaches. If you would like to help, it says, then contact David Shackleton, giving both the telephone number and the address of the house behind.

When he comes to the door, I explain to David that, while I can't help physically, at least not more than a passing sweep, I am interested and might be able to support his efforts with whatever publicity comes from the eventual book. Over a cup of tea, with the incessant rain beating against the windows of his cabin-like front room, he explains how it started.

'I was an RSPB lifer,'[15] he says. 'When I retired from being the warden at Haweswater three years ago, I realised how much I minded about the plastic. I saw it as a real and immediate threat to wildlife, a real pollutant and a real eyesore, and I saw that fishing waste was by far the biggest element of the problem. I had the time, and I wanted to get something done. Then maybe if others saw me, they would do the same.'

As so often happens, what started as a hobby developed to be an obsession, 'a pilgrimage', as David calls it. 'People who care are in the minority,' he continues, acknowledging that it is less easy to care when you are not sure where your next meal is coming from. 'The challenge is to engage with everybody, and not just the already engaged. Fishermen never come this way to see the mess they leave and elections change nothing, so we need to find our own ways of making things better.'

The room is lined with old bird books that I find I have an irrational desire to read. He could say nothing at all and his walls would still eloquently explain just who he is and why he does what he does.

'The garden idea started as me just wanting to shock people. But, by keeping it neat and tidy, it became an accepted feature of the village. I think people are quite proud of it these days. Locally, it's quite well known, but this place isn't really on the tourist trail, so news of it doesn't spread that far.'

Much of what lies in his garden comes from a spectacular one-year haul he took from Annaside Beach, a few miles south, with occasional help from friends. He rummages around for a

---

[15] Royal Society for the Protection of Birds.

magazine article that covered this work, and then reads out some of the recovered quantities.

'Twenty dump bags of fishing waste; 30 plastic fish boxes; 4,000 plastic bottles, 3 large plastic barriers; 22 safety helmets; 30 dump bags of fishing rope; 35 tyres, including 2 tractor-sized; 2 fishboxes of lubrication tubes; 650 items of footwear; 290 cigarette lighters; 12 bicycle saddles; 6 fishing rods; 11 pram wheels.' He looks up to see if I am writing it down, to see if I want more. 'Two fish boxes of assorted tennis balls; 4 dump bags of agricultural plastic.'

'One beach,' he says. 'One year.' You can spin it any way you like, we agree, but this is the evidence of a society utterly ensnared by its obsession with owning stuff solely for the hell of owning it, and with growth for growth's sake.

I don't ask him my normal question about whether he is a pessimist or an optimist, as I find that I don't want to hear the answer. He senses it and replies anyway.

'We're probably screwed,' he says. 'But at least I'm doing something about it.'

ART, ACTIVISM, AND A BRIEF INTRODUCTION TO NATURE CONFLICT 5

Morecambe Bay. Late July

> 'Knowing when to fight is as important as knowing how'
> Terry Goodkind, *Faith of the Fallen*

At first, and for only the merest suggestion of an instant, I get the impression that someone is drowning themselves.

Later, I learn that's what people often say, at certain states of the tide, when they first set eyes on Antony Gormley's mass installation *Another Place* on the beach at Crosby. Whether in the stillness of a summer's day, or in the bitter wind of an early February gale, you come over the dunes behind the beach and suddenly see a hundred human-sized figures up to their ankles, waists or necks in Morecambe Bay, which comes as a strange and moving sight. No piece is particularly close to another one so, rather than a community, I feel a powerful sense of aloneness coming through. As someone who frequently struggles to appreciate modern art, either through failing to understand the skill involved or through irritation at the self-referential language so many artists use to describe their own work, this strange fraternity appears to speak to me without me needing to understand it. I'm surprisingly grateful when art does that. It reminds me of Michelangelo's comment about his own work: 'I saw an angel in the marble and carved until I set him free'. On those rare, perfect days, writing can be a bit like that.

Art, as such, had not been high on my philistine agenda when I first set out, but it has gradually promoted itself through a series of unlikely bedfellows: a sculpture of stones below the tideline at Sandwood Bay, a Celtic cross by the abbey at Iona, a seascape in oils in a tiny coffee shop in Arisaig, a child's painted pebble for sale in the driveway of a house on Arran. Bit by bit, I find that I start to make small deviations from my route to see cultural things, and bit by bit, the art I see helps me to understand a little better my relationship with the ever-changing blue body of water to my right, even if only subliminally, without my ever knowing how. It is almost as if art is the inevitable result of all that light and those huge skies meeting 5,000 years of written history. As I progress onwards around the island, it will often be art that creates the polite clearing of a throat on the occasions when nature falls silent: graffiti on some harbour wall; an incomprehensible installation in Littlehampton; even a Christmas tree in Cornwall, made entirely of redundant fishing plastic. If art is an expression of our collective memory, as people sometimes say it is, maybe *Another Place* takes us back to the heart of what it is to be an islander.

Over half an hour, I visit maybe twenty of the figures, all based on a body cast of Gormley himself, and all staring enigmatically out towards the Burbo Bank Offshore Windfarm in the Irish Sea. The sea is too deep, or the bed too rough, for anywhere north of here to be suitable for wind turbines, but there are twelve farms here in Morecambe Bay and more to come. Just out there on the western horizon are two of the six 'Round 4' projects for helping the UK meet its target of generating 50GW of offshore wind energy by 2030, one just off the coast where I am and the other further north off Barrow-in-Furness. It is difficult to exaggerate just how much the government is pinning on offshore wind to help it meet its 2050 net zero target (after all, 50 GW is sufficient to power around half the homes in Britain),[1] and just how hard it will be to achieve in practice.

One of the figures even has a ripped blue Everton strip flapping listlessly around its torso, another a washed-out white tee shirt. After half an hour or so of wandering around, I am as far out in the water as my wellies will allow when I realise that the ground below me is soft, and it is time to head back in.[2] As I head quickly inland, a large elderly man is heading equally quickly seaward in my direction, running, not in the sense of covering the ground quickly, but in the sense of all his body parts moving as quick as age and girth will allow them to. I suppose that he is a local, about to berate me for having gone further out than I should, and I decide to take whatever he says on the chin.

It turns out that he just wants to know if I have seen a mobile phone that he has lost, which I have. As I walk with him towards the dune to retrieve it, I ask him what he makes of the installation.

'Dunno, really,' he says, and then tells me that he comes down here to see them every morning of his life.

In many ways the Dee Estuary, which marks the northern point of the Anglo–Welsh border and is where my journey turns westwards, is uniquely useless to our modern, capitalist lives.

This is because a relatively small amount of water occupies a very large area, meaning that, when added to the accumulations of silt that spills down from the hills of mid Wales, it is virtually unnavigable. While the waters of the nearby Mersey were creating a deep navigation route over the centuries, on whose banks the port

---

[1] On which target no less than £1 trillion of revenue for British businesses and 480,000 jobs are within reach, according to *Mission Zero, Independent Report on Net Zero* by Rt Hon Chris Skidmore MP.

[2] On the evening of 5 February 2004, at least 21 Chinese illegal immigrants were cut off and then drowned by the incoming tide in Morecambe Bay, while harvesting cockles. Morecambe Bay is not to be messed with.

city of Liverpool grew, the Dee Estuary was simply scoured out by a large glacier to the north, and always remained shallow. This, while being bad news for entrepreneurs, ferrymen and traders, has been a huge blessing for wading birds, for whom the estuary's shifting sands provide one of the finest winter-feeding grounds in Britain. Huge flocks of up to 10,000 knot and dunlin rub shoulders with scoters, scaup, sandpipers, short-eared owls and any amount of local rarities, who all happen to have been brought here to feed by the side effects of nature's most unstoppable of powers, the tide.

Historians will rightly tell you that our island ancestors were generally slower off the mark in life sciences than their distant Roman and Greek cousins around the Mediterranean, but not as far as the tides were concerned. This may have something to do with the average tidal range in Britain being a full seven metres, which made knowing about it, and predicting it, a matter of life and death. Around Venice, in contrast, it is under a tenth of that, and in Athens, near where Aristotle is said to have drowned himself in frustration at not being able to understand how tides work, it is less than ten centimetres. The first known tide table was written in around 1220 for London Bridge, and if it's a safe bet that mariners understood its relationship to the gravitational pull of the moon a thousand years before that, it's an even safer one to say that birds have been flocking to the wet sands of an ebb tide since they had wings to fly, and that birdwatchers have quietly followed them there to marvel at the sheer bounty of nature on display; it is far from rare to have 50,000 knot swirling around here on the best days. Indeed, the waders shape their lives around it, and all of them have evolved to some degree to take specialist advantage of a particular foot type in the mud.

You can normally tell the primary food choice of a bird by the shape and size of its bill, varying from the little grey plover who will forage around on the surface of the mud for cockles and snails, to the fifteen-centimetre de-curved curlew beak, armed with hundreds of sensors to probe for worms far below the surface. While we might see the tide as a predictable force of nature, waders see it as a giant, permanently restocked larder. And while waders treat it as a life force, we humans occasionally treat it as yet another reason to have a good argument.

On the other side of the River Mersey, I find I like Hoylake from the start. I like the slightly rakish faded grandeur of a town that seems happy on the face of it just to be taken as it is; I like the cleaning ozone breeze that scuds into its streets and alleyways across the open sands from the Irish Sea; I like the deceptively peaceful way it is going about its business. I like that it has rolled with the punches of its recent history, the loss of its port, its fishing industry and its fashionable sea bathing. But most of all, I like the town's 'wader wall', a delightful stretch of wall along the sea front; on each of its 40 panels a different local artist has painted individual waders, or groups of them. Taken as a whole, with its terns, oystercatchers, sandpipers, knot and curlew, it is a free and ever-present resource that surely delights people going about their daily business. Particularly children. Nature education needs all the help it can get.

In stark contrast to the trend elsewhere, the sea is actually retreating from Hoylake, and doing so at the rate of around six or seven metres a year,[3] as sand that has eroded from the sea cliffs of North Wales is gradually blown inshore from the vast sandbanks near the mouth of the Dee, in a process known as saltation. Not only is the sea retreating, but the beach itself is rising at ten times the rate of the sea level, to the extent that the whole thing rose by 600 millimetres just between 1980 and 2000.[4] These days, the windows of the Hoylake Sailing Club stare enigmatically at the flat sands stretching out to a now distant sea that as recently as the 1950s lapped at its walls; close by, the lifeboat station, one of the oldest in

---

[3] Coincidentally, the same distance that the sea is grabbing from the cliffs 200 miles away on the Yorkshire coast.

[4] 'The beaches at West Kirby and Hoylake: Options for managing windblown sand and habitat change'. Paper by Wirral Council 2000.

England, now has to employ the mother of all caterpillar tractors, a jet boat and a hovercraft to get help across the sands and to the water beyond. This all means that Hoylake is technically no longer so much a village on the sea as a village a good mile away from it, and tiny mobile dunes are appearing below the sea wall, where the high tide once lapped. A little bit of this is Atlantic salt meadow, and these 'embryo dunes' are the forerunners of bigger dunes that will come in time. They bring in with them a whole new botany, including Sea Milkwort, Strawberry Clover, Scurvy Grass and Sea Aster. For the time being this also includes, for its sins, a grass called *Spartina angelica*, an invasive species that originally arrived in the ballast water of ships coming in from North America in the 1870s. So far, so simple: coastlines and habitats change, and this place is changing. Besides, salt marshes take carbon dioxide out of the atmosphere, which is more than bare sand does.

But since the 1960s it has been this grass, what it looks like, how it behaves and what to do with it, that has driven a strange wedge between two communities, dividing those who would like a clean leisure 'amenity' and those who would like to protect natural succession in a world that they feel has precious little of it left. Spartina, although it is now just one of hundreds of botanical species on the beach, is the one that gets the blame, the one always there in the 'Wanted' photograph. Different measures have been used over the years to control it around the country – raking, digging, poisoning and rotoburying, for example[5] – and the evidence seems to suggest that there is no one solution that works everywhere, that it seems to have run its course in the south but is still spreading in the north. After all, it is not as if there aren't thousands of acres of 'clean' sand everywhere else as far the eye can see, it is just they are not precisely where the ward councillors would like to see them, close to the Victorian promenade.

---

[5] A technique that sifts out the spartina roots and suffocates them by burying them in a deep trench.

To be fair to those who yearn for pristine beaches, the green embryo dunes are not what passes for picture-postcard pretty these days, and you can equally see why local people might grieve the loss of a nice beach, particularly if they have already lost the sea that once washed over it. But it is a tiny area relative to the surrounding sand, and the fact is that all habitats are dynamic, not frozen in aspic, and the new dunes form an already evolving ecosystem with inevitable winners and losers. Surely humans are allowed to be small-time losers from time to time. Besides, there is a precedent. This happened in Lytham St Annes up the coast a century or more ago, and now it is winning awards for its picturesque, sandy dunes.

For years, the beach was regularly treated with glyphosate (a herbicide, it is worth mentioning, that is banned in 33 countries including Canada, France and Germany) and then raked off at the behest of the council. Then, after the 2019 treatment, the tables turned when newspaper columnist George Monbiot and TV *Dragon* Deborah Meaden joined the campaign to let natural succession have its way. Natural England, never particularly keen to enter fights, came perilously close to entering this one, putting out a statement in August 2020 to the effect that there was only one more year left on the current spraying licence, that it was extremely unlikely to be renewed, and that Wirral Council would do well to remember their legal duties under Section 28G of the Wildlife and Countryside Act of 1981. For now, there is probably not a cat's chance in hell that permission would ever be granted again for the wholesale removal of grass for this clean amenity beach. You can produce an alphabet soup of initials that protect its current status, from LNR, SSSI, SPA, SAC and RAMSAR.[6] Somewhere down the line there may even be

---

[6] Local Nature Reserve, Site of Special Scientific Interest; Special Protection Area; Special Area of Conservation. Ramsar Convention, named after the Iranian city in which it was signed in 1971, is an international treaty for the conservation and sustainable use of wetlands.

a compromise to be had, although it seems on the face of it hard to know where that would come from. But the fight is far from over.

On a windy afternoon, I wander out through the embryo dunes with Jane Turner, an ornithologist, retired biophysicist and passionate activist for the biodiversity that the retreating seas have accidentally gifted. 'Pioneer salt march is one of the rarest habitats in the UK,' she explains. 'It is marsh, but what makes it so rare and special is that it is *transitional* embryonic dunes and *transient* Atlantic salt meadow that is accreting sand and not silt. It makes up about 5 per cent of the entire UK stock of the habitat. Back in the 1950s, people were convinced that spartina would spread like a triffid, as it had previous form of exploding into formerly unoccupied ecological niches. These days, UK conservation bodies are more relaxed and even supportive of spartina in all but exceptional circumstances. What is going on in Hoylake now is pretty similar to the fight against ragwort. Based on bad science and scare stories.'

I press her on what biodiversity gains have been made since the retreat of the sea. 'A great deal,' she says. 'But I'll offer two. Linnet were almost extinct on the north-west Wirral twenty years ago, and now the post-breeding flock amounts to around 130. We have also got twenty of the less than 300 shore dock plants in the whole country, where the UK is the most important population in the world.' Josh Styles, a local botanist, has identified no fewer than 240 plant types on the beach, of which 27 are at 'the risk of extinction, regionally or nationally'.[7]

The same wind that is turning the huge blades of the turbines of the Burbo Bank Offshore Wind Farm five miles to our north is blowing Jane's hair across her face, giving her an almost biblical campaigner's look against the flat, dun-coloured backdrop of the Wirral sand. When I ask her whether the beach lovers have a point, whether the vegetation will spread and spread like they fear, she is

---

[7] *Birkenhead News* article, 23 Feb 2023

resolute in her reply. 'The Hoylake area is not getting bigger,' she says. 'Satellite imaging shows that it is staying the same size and actually reducing as a percentage of the growing intertidal beach, but that what it *is* doing is getting greener and greener, and that is because there is more and more variety in there every year.'

As we walk back to the promenade, she tells me that she understands some of the locals *wanting* clean beaches, but that wanting something doesn't make it practically possible, any more than she could get her twenty-year-old knees back. 'Our shoreline is moving,' she says, 'so getting access to the new shoreline is where our energy should be directed.'

Hoylake's ward councillor, Andrew Gardner, a leading and highly vocal light of the 'clean beach' campaign, accepted my offer of putting his own side's view with alacrity, emailing the following:

> There is no local argument. The people of Hoylake overwhelmingly want their beach back.[8] They only lost it for the undue influence of some eco zealots and a CiC interested in a financial grant.
>
> There are no dunes be they embryonic, fledgling, proto, pseudo or any other phrase the lunatic fringes like to claim.[9] It's four years in and the beach level is flat if higher because of trapped sand in vegetation that survives because of promenade drain run off not because of natural accretion. The whole counter argument is evidently a fantasy.

---

[8] Er, not that overwhelmingly. In a local consultation, 45.3 per cent wanted 'sand dunes to develop across the entire area', 37.3 per cent thought that 'some sand dunes were acceptable' and only 17.4 per cent didn't want sand dunes at all: https://haveyoursay.wirral.gov.uk/hoylake-beach-information

[9] Many local academics, two of whom are dune and dune system specialists, seem to think they are dunes.

You are right in the human need though and Hoylake beach was designed to be Hoylake's equivalent of a park for recreation. It's a betrayal of democratic justice to sacrifice open public space for the rat and mosquito infested dog poop scoop that my residents currently have to endure.[10, 11, 12]

I hope I have made my position clear!

Abundantly so. When I ask him afterwards if he thinks that there is a possible compromise, his reply is no more optimistic: 'The problem that the eco lobby has is that this has been a heist, a smash and grab that was potentially illegal. The majority of the residents will not be forgiving. They have been collectively abused.'

I'd never thought of that before: natural succession as some sort of smash and grab raid on humanity.

It is not as if I have a dog in the fight, but there is something ugly going on in the dark corners of this local spat, something that includes highly personal vilification of respected academics and campaigners, and therefore something that keeps me digging. There are arguments on both sides, of course there are, but science that is selectively and misleadingly deployed is an ugly weapon, and many is the doormat that has received leafleted scare stories about methane from declining salt marshes, or wading birds disappearing entirely from the new vegetation. Most of us like our conservation debates to be apolitical, where they can be, but what is remarkable is the almost comical aggression of the leaders of

---

[10] Rats avoid sandy substrates as their burrows collapse.

[11] Liverpool academic (and world expert on coastal mosquitos) Professor Matthew Bayliss assesses that there is no chance of a mosquito issue at Hoylake.

[12] The implication being that discerning dogs would only poop on the grass, not pure and pretty sand.

the 'clean beaches' campaign. 'When listening to, and voting for, so-called environmentalists in the town hall,' runs one rant on a semi-official Facebook page,[13] 'be aware they back extreme views. Remember, you're choosing extremists.' They follow this up by rather disingenuously illustrating the post with pictures of Insulate Britain campaigners defacing targets in London over the use of fossil fuels. A separate post on the more radical 'Save Hoylake Beach' Facebook page refers in cheerful Soviet-era terms to 'brain-fried Swampies with their pathetic political supporters' and 'God only knows what their home toilets are like, but maybe that's why they wear ill-fitting tee shirts and old, smelly clothes'.[14] Josh Styles, the botanist who has been studying the beach for years, says that he has absorbed serious abuse, which he puts down to being on the 'wrong' side: 'I've been called a "swamp monkey"', and told I'm not credible because I'm autistic. I've also had people giving me minor homophobic abuse.'[15]

We all have a duty to be as robust as we can, these days, but I think this matters. Respectful argument is a vital element to arriving at solutions that are as right as possible for human communities and for nature, especially when they are not seen as mutually inclusive. Both sides are obliged to listen and, generally, both sides are obliged to compromise. What no one should have to do, whatever side they are on, is absorb intimidation and abuse for stating their beliefs.

Beyond these niceties, the important debate, however, seems to me to be one between anthropogenic man and embattled nature, with the former demanding to live a life whose primacy is utterly unencumbered by the needs of the latter. At Hoylake, there is one choice that is visually right for a portion of just one solitary species

---

[13] Ward Councillor News for Hoylake, Meols and Central West Kirby.
[14] Save Hoylake Beach. Post from 31 Jan 2023.
[15] *Liverpool Echo* article, 12 April 2022.

and another that is practically right for almost everything else, and yet the debate is still on a knife edge. In this most de-natured of countries, it is not an easy debate in which an environmentalist can be neutral.

Well might we like nice, clean beaches conveniently nearby, but we need the biodiversity more. All of us do, irrespective of whether we kid ourselves that we don't.

The Priest

Bardsey. An island in the tides.

The puffins are long gone from Bardsey Island when I get there, and only the shearwaters remain, slicing their busy way over waters that are 'as sleek as seal pelts'.[16] The foghorn moaning sounds of the grey seals is there, too.

Bardsey – or Ynys Enlli, to give it its more poetic local name – is the northern gatehouse to Cardigan Bay, and probably the most quintessential island of the 30 or so that I will visit on my journey. It has no bridge to it, unlike Skye, no estate agents, unlike Mull, no airstrip, unlike Benbecula; it is not part of an archipelago, as the Outer Hebrides are, nor is it within apparent spitting distance of the shore, as Handa or Skomer are. Above all, it is always challenging to get to.[17] It doesn't even have electricity to service its miniscule population. Instead, under its whaleback form, which broods a long couple of miles across the treacherous Bardsey Sound, are reputed to lie the bones of 20,000 saints. 'In every part of the island which is free of stone,' complained poet and artist Brenda Chamberlain, who lived there for fifteen years after the Second World War, 'the spade strikes against human thigh and breast-bone.' She went on to describe it as at once sanctuary and prison: 'This is a land that hoards its past, and merges all of time in the present.'[18] I quickly see what she means.

In Celtic lore, Bardsey is known as a 'thin place', where the veil between this world and the other is at its slenderest, and

---

[16] *The Turning Tide*, Jon Gower.

[17] When I was writing a previous book, *Shearwater*, a combination of weather and Covid resisted no less than four attempts I made to get to the island.

[18] *Tide-Race*, Brenda Chamberlain. Seren Books, 1962.

in the quiet of its abbey graveyard, I find myself idly wondering what exactly it is about islands and faith. From Anglesey, Iona, Lindisfarne and Lewis to a hundred isles beyond, some form of faith had made its home there long before I came poking round. I reflect that if I'd had a penny for every candle I had lit over the years in some island church in my hopeful agnosticism, I'd have the start of a small fortune. Those pioneering pilgrim monks from Ireland, those coffin roads on Eigg and Barra, those pilgrims making their way towards Iona, those long-ago hermit monks on Inner Farne Island, they all articulate some sense that if you are ever to find your god, an island is a pretty promising place to start. As one writer put it: 'Until the Reformation, Britain was ringed by a forcefield of monk-inhabited islets',[19] and it seems that those early ships and coracles of St Brendan and St Columbus still cast a long shadow over our outer fringes. I am still mulling this over a little later, approaching the tiny island shop with its honesty box and its esoteric selection of Pringles, postcards and dried pasta, when I am greeted by a large, beaming man in green dungarees and a purple shirt, sitting at a table and surrounded by his own art, sun reflecting off the tiny bottles of ink around him.

'Go on, please take one,' he says, gesturing to a pile of little paintings framed in white card. 'It's what I do. Paint things to give away.' It seems an unusual way to make a living on the face of it, so I cannot help but fall to chatting with him, helping myself to a small bright blue seascape in the process.

It turns out that Chris Duffett is in rather a good position to provide a few answers to my question. He is on the island each year for two weeks, part of an ecumenical rota of chaplains who look after the spiritual health of the place through each summer, ministering mainly to the residents of the nine holiday cottages

---

[19] *The Britannias*, Alice Albinia. A good place to start a voyage of discovery on this topic.

that make up most of the transient population. He is a Baptist, his is a big-hearted God, and his is a gospel of simple words. His healing work, he tells me, more normally takes him to the Wetherspoons pubs of Cambridge and Peterborough. 'It's where I am most needed,' he says with a cheerful shrug. 'Chatting to dispirited Amazon night-shift workers whose release valve is found in cheap, mid-morning alcohol.'

Each evening, here among the wheeling gulls of Bardsey, he hosts an informal healing service and is always surprised by how much better attended they are than they might be on the mainland. If I were here for longer, I think to myself, I would surely go along, for the coast works its mysteries on me as well. Sometimes, you simply don't need to know why.

'There's a strange tension,' he says, 'between Christendom and post-Christendom on the island. Most people will happily engage with the chaplaincy even if not particularly religious. Even if not religious at all, in fact.' A couple of day visitors wander up to see what is going on, for an instant they appear traumatised by the offer of something irrationally free. 'Go on,' he says again. 'It's what I do.' They look briefly to me, seem reassured that I have survived the experience, and take one.

He makes no wild claims; if there is healing, he asserts, it comes from within, and maybe from the island. Whatever your faith, he says, there is surely room amid the anxieties of our modern lives for a bit of spiritualism. There is enough hard graft and stern practicality in our everyday lives to allow for something gentler, less pervasive, to make itself felt within us.

'Sometimes, we just don't need to analyse and understand everything that happens to us,' he says. 'A little bit of mystery doesn't do any harm.'

He is right. My coastal odyssey may be a long one, but through it all I will never lose that sense of otherness that an island gifts me.

That and the uncertainty of the return boat.

THE LOST KINGDOM OF CANTRE'R GWAELOD: CARDIGAN BAY

6

Mid September

> 'Ships are expendable; the whales are not.'
> Paul Watson. North American environmental activist.

From Bardsey Island (Ynys Enlli) eastwards and then south around Cardigan Bay, the nature of my journey changes suddenly, and completely.

Up until this point, in working my haphazard and untidy way south from Cape Wrath, now 400 miles behind me, I have always thought of myself as travelling down the western edge of the British landmass, jumping from island to mainland and back again, but always referring back to that 80,000 square mile lump of land over my left shoulder. From now on, starting in the tiny former fishing village of Aberdaron and continuing all the way to Land's End, I find instead that I am mentally on the eastern rim of a huge Celtic sea whose deep connections come from shared history and culture, and from an ever-moving network of seabirds and cetaceans in between.[1] This is not so much a matter of distance (although

---

[1] The Isle of Man, to the north, is very much part of this Celtic notion, where heroic efforts are simultaneously being made both to restore the Manx language and to continue to help wealthy people to avoid taxes.

Holyhead is four times further from London than it is from Dublin) as of attitude, language and, above all, culture. It is as if the whole place moves to the beat of an entirely different drum to the one with which I am familiar in my Sussex home. It is a cultural pull that I won't experience again until I am at the very end of my journey in Caithness, where the pulse that runs through the place is sometimes decidedly Nordic.

It helps this sense of Celtic connection that Cardigan Bay is a classic convex shape, where, on a clear day, you can see Yr Wyddfa[2] all the way from Skomer, and Bardsey from 40 miles away in Aberystwyth. In fact, if you stand on Mynydd Enlli, the highest point on Bardsey, you are in easy sight of the Wicklow Mountains just south of Dublin. For some reason, I will feel that centripetal pull towards the middle of the Irish Sea until I have rounded Land's End, and started heading eastwards.

Walking my way round various different sections of its coast path, on the Llyn Peninsula and in Ceredigion and Pembrokeshire, I feel myself more involved with the sea beyond the shoreline than I ever have before or will again. The land to my left has gone from being the anchoring solidity of my journey to a mere reference point. At first, I put the new feeling of change down to the autumnal equinox, with its regular horizontal bucketing rain that makes a mockery of my waterproof jacket and trousers, but slowly, village after village and headland after headland, I come to understand that the changing sense of place has once again captured me. I walk past overhanging hedges fat with hawthorn berries and the first sloes of the year, opposite rainbows, above pupping seals and under skies of gulls and aerobatic choughs; I wander past doomed sea defences, redundant fishing boats and hillside cities of static caravans. Cliffs give way to long shingle beaches, then sandy ones and back again to cliffs, and I mark off the miles with pretty sea

---

[2] The mountain formerly known as Snowdon.

villages, all seemingly with 'no vacancy' signs on their bed and breakfast cottages. Half-closing my eyes, I sometimes feel that this could be the north coast of Cornwall, only without the crowds, the spotless SUVs and the eye-watering prices.

But Cornwall it is not. Welsh is the first language here, insistent if not defiant, just as Catalan is in Barcelona. There is a palpable sense not just of separation from London, but also from Cardiff, nowhere more than in the battle to restore nature to the coast. When, in 2018, a regional initiative called Summit to the Sea was deemed to be dropped patronisingly into planning by mainly English academics and campaigners, its first incarnation was seen off by the locals in no uncertain terms for being considered to consist of three toxic elements: outsiders, arbitrary lines on maps and the relatively alien concept of rewilding.[3] As on Mull back in June, so here in Ceredigion: there is a palpable sense of how good things could be if only it were locals making the decisions.

Locals like Joe, for example.

I wouldn't go so far as to say that Joe Wilkins represents the whole future of Welsh and British nature restoration, but I would think that he is very likely to be somewhere in the vanguard of people who are. He is part of a new breed of highly qualified, highly motivated twenty-somethings who are no longer content to be just the humble volunteers on beach cleans and in bird reserves, but are actively involved in shaping policy, too. He has been until recently, among other things, Head of Campaigns at UK Youth for Nature, an impressive movement calling for urgent action on the crisis in nature. Unlike some similar NGOs, the website is refreshingly free of feelgood platitudes and buzzwords, and full of a determination to

---

[3] For a good description of this, you might read the penultimate chapter of David Elias' excellent account of a Welsh hill farm, *Shaping the Wild*.

be a political force for young people. At 24 years old, he is already a veteran of many meetings in Westminster and Cardiff. He avoids the easy trap of thinking that all politicians represent the problem, not the solution, and is clearly grateful for the respectful ear that the young Plaid Cymru MP for Ceredigion, Ben Lake, has consistently given him.

Small 'p' politics is important to him, especially in Wales, where the imaginative Wellbeing of Future Generations Act was passed by the devolved government in 2015 to try to ensure that all public bodies put long-term sustainability at the forefront of their thinking.

'It's a start. And at least the Welsh government is paying nature lip service. But you'll note that the future generations they talk about are only human ones,' he explains as we walk along the coast path to Aberystwyth. 'It would have been good to have included all the non-human species in the act, as well, as they are the ones getting shafted.'

These days, he is well used to being the youngest in any given room by as much as 30 or 40 years, but he no longer cares, as he feels he has a golden bullet of an answer for most things. Long evenings in stuffy meeting rooms with his mainly male elders and betters telling him that such and such can't be changed as surely he understands that it has 'always happened this way' simply gives him the opportunity to ask: 'Yes, but has it worked?' He has form in this approach, as the thesis for his Master's programme was on how to get young people meaningfully involved in conservation. Like me, he is no fan of the overused word 'passionate', but there is something almost visceral in the depth of his ambition for nature and lack of ambition for himself that suggests that he is, indeed, passionate. Generally, in my experience, it is the ones who say they are that aren't.

Certainly, the depths of his roots in the Ceredigion coast are seldom hidden for long. He breaks from our conversation on policy to point to the broiling sea beyond one of the long-submerged causeways and starts to tell me about the legend of Cantre'r Gwaelod. 'Somewhere out there,' he says, 'there was once a network of prosperous cities and farms protected by a series of dykes, all of which had to be closed at night. That's where those causeways lead. Then it's the normal story: one night the royal gatekeeper gets drunk and forgets to close his sluice gates; the water comes in and inundates the land, and that is the end of everything.'

Legends, I think, sound so much better when told in the dialect of the local coast. Most seem to have some small element of historical truth to them or are at least based in part on a cautionary tale. I ask him if that is the case with Cantre'r Gwaelod.

'Maybe,' he says. 'Maybe not. After all, the sea level has risen around 30 metres since the end of the last ice age, so there certainly would have been more land back then than there is now. Old tree stumps and other bits of submerged forest come up from time to time, and occasional lumps that could have been part of some structure. But I think legends have always been associated with warnings, and this was maybe a warning about what was coming down the line with the rising sea.'

If it's a warning, it's a quite a timely one. Just behind us is the village of Borth, one of two on this coastline that is only being kept above the waves by the deployment of expensive new sea defences. The other one, Fairbourne, allegedly lacking the powerful middle-class voice and resources of its neighbour to the south, will gradually be returned to being a tidal salt marsh, with at least 420 houses being demolished, along with roads, sewers and gas and electricity infrastructure. Some 850 residents, Britain's first 'climate refugees', will have to find somewhere else further inland and drier in which to live. In undergoing this fate sometime in the next twenty years, Fairbourne will become the first village in the United Kingdom to be 'decommissioned' because of climate change. Quite ironic, really, when it is not even on the side of the country that is actually sinking.

We are heading south to where a friend of his works at the Cardigan Bay Marine Wildlife Centre in New Quay; in a couple of days' time, the questions from me will be about dolphins, porpoises and whales. It is the day that Prime Minister Sunak has announced a watering down of some of the key commitments on the route to net zero, and half of Joe's attention is caught up in the production of memes that the charity will be putting out later on in the evening, in wide-eyed reaction.

He talks about the sea life, about how the bay holds the UK's most important population of bottlenose dolphins. 'They're not as sweet as you would like to think,' he laughs. 'Would I knowingly go out swimming if there was a mother and calf in the bay? Would I hell. Would I even go paddle-boarding? Probably not. You only have to see a bottlenose taking out a smaller harbour porpoise and you know that you are in the presence of a proper apex predator, and a large one as well.'

We walk on. To give myself a little welcome breathing space on a long and steep ascent, I ask him a question calculated to get a pleasingly long and detailed answer, long enough at any rate for me to get my breath back. 'If you could only influence one area of policy, just one thing that you can move forward appreciably, what would it be?'

He stops and looks down at the rip tide boiling up on one of the causeways out in the bay, one of those fabled roads into Cantre'r Gwaelod. 'I'd simply want to change the language of conservation,' he says. 'Basically, all the aims now are only for some form of damage limitation. They want to slow down the decline of things, protect a few endangered species and all that, but there is not yet any sense of a determination to bring bio-abundance back. Doing as we have always done will only give us what we have always had. Meaning decline.'

After a bit, he stops and looks me squarely in the eye. 'My goal is nature recovery, pure and simple,' he says. 'That's all I want to do with the rest of my life.' The eyes ask me to write that down, even if the mouth doesn't.

For now, Wales and Britain are lucky that Joe is able to live at home with his parents, for it partially disguises the mainly pitiful value we as a society put on the cutting-edge work of skilled environmentalists like Joe as opposed, say, to computer programmers or currency traders. He could do many other local jobs tomorrow for more money than he gets in his portfolio of activism, and it is our great good fortune as a society that he and others like him choose not to.

A strong and wet onshore gale reminds us that it is the day of the autumnal equinox when we arrive at the Cardigan Bay Marine Wildlife Centre in New Quay a couple of days later. When each rain shower homes in on the harbour to bookend the periods of bright sunshine, it is of the kind that renders pointless almost any attempt to keep dry.

Before breakfast, and before the rain sets in, I go for a walk southwards from the main harbour. All along the little seafront road, small groups of residents are out in their front gardens, often in dressing gowns, sometimes with binoculars and generally with mugs of tea in hand. It is an immutable law of nature that when two or more people are gathered together looking in the same direction, it is worth stopping to see what they are looking at. In this case, there is a group of five bottlenose dolphins swimming in what looks like slow circles a short way off the coast. It strikes me with a comfortable feeling that no one here is in a hurry. Not the dolphins, not us. They stay, we stay. New Quay happens to be one of the best places in Britain to see these creatures from the shoreline,[4] which is just as well on a day when conditions make it highly unlikely anyone at the marine wildlife centre will be taking a boat out.

The centre was established in 1996 by local marine wildlife guide Steve Hartley as a vehicle to study in depth the mammal sea

---

[4] Scottish friends and colleagues assert the quality of Chanonry Point on the Moray Firth as a spot for watching cetaceans, so you can take your pick.

life in the wider bay area, especially the bottlenose dolphin, for whom the area represents the most important habitat in Britain, and through that, to help conserve what was left. Since then, the growing team of scientists, interns and volunteers have created a vast information bank of the cetaceans in the bay, not least in their ability to identify at least 200 individuals from the possible 250 bottlenose dolphins by the marks, nicks and scars on each one's fins. Contrary to what many people might imagine, much of conservation science comes not from adventures out on the high seas, but from hour after hour of observation, and of inputting information into computers which can then be shared with the wider academic community. If you have ever visited New Quay and seen a motionless figure in a yellow high-visibility jacket standing on the sea wall and staring out to sea for hours on end through a pair of binoculars, you will have seen one of their volunteers in action, part of an unbroken citizen science project that predates our millennium. Dolphins come close to shore and, day after day, for those 27 years and counting, their visits and behaviour are recorded. Just as important as the research is the outreach work they do to engage the public, especially children.

The office is energised by a constant flow of information. In the corner, a volunteer is inputting the stream of data that comes in from boat trips; next to him, someone is working on dolphin acoustics; and on a big screen opposite, another volunteer is painstakingly identifying dolphins seen recently against the library images of the whole population. 'Here you go!' she says triumphantly, when I'm looking over her shoulder. 'Thick nick halfway down the back of the fin, and four rake marks below that. That's a match.'

Maddy de Marchis has been a marine conservation intern for the last year, having been a visiting seasonal volunteer before that. A career in the environmental world needs qualifications these days, so she is also studying for a degree in Environmental Science at the Open University. Her colleague Laura Evans has been there since 2015 and is the Marine Project Officer. In truth, everyone round here does a bit of everything. My own meagre contribution is a packet of chocolate digestives to fuel the discussion. The cupboards of wildlife establishments around the whole coast of Britain are probably populated by the stale remnants of biscuits I bring with me, mainly for my own consumption.

I start provocatively by asking why we need cetaceans at all.

'The bottlenose dolphin is an apex predator at the top of the local food chain,' says Laura. 'Remove it, and you remove the building blocks for everything underneath. That's how trophic cascades work. Like in Alaska where they killed all the sea otters who then didn't eat the sea urchins who then multiplied and ate the kelp until there was no life left on the shoreline. Including sea urchins. The bottlenose is lucky in that it is a generalist feeder, and will eat more or less anything: salmon, seabass, mackerel and trout, for example. But it will also eat squid if it has to, and it has the ability to travel big distances, up to 100 miles, to get a good meal. They are an umbrella species, in that if you protect them at the head of the food chain, everything else benefits. Porpoises are less lucky, as they tend to be specialist feeders.'

Recent history teaches us that the Anthropocene is not a good time to be alive if you happen to be a specialist feeder, especially if that specialty is sand eels. Fundamentally, the bottlenose is doing reasonably well, but the picky porpoise less so.

I pass the digestives around, but only Laura and I take one. I ask what the biggest threat is that the dolphins currently face. 'It has to be disturbance,' says Maddy. 'Disturbance by humans. Not so much by the wildlife boats, as they operate under a code of conduct and generally behave respectfully. It's more just the endless coming and going of boats from the coast. Fishing boats. And anything with an outboard motor.'

Not a single person I meet in my year on the coast likes jet skis, but when I suggest that they must be one of the main culprits, Joe stops me short. 'Not really, as there aren't that many round here.

They would be if they could. No, funnily enough, the biggest problems can sometimes be the kayakers and paddle boarders. Sure, they don't have engines throbbing away under the water, but that fact seems to make them feel entitled to go that much closer and to cause genuine disturbance. Particularly with mothers and calves. It's just needlessly putting them into a place of conflict.' Joe is a keen paddle boarder himself and spends a lot of his time gently persuading people to back off.

'The worst is with seal pups,' Joe goes on. 'Paddle boarders taking selfies near a defenceless pup probably aren't aware that they could be pushing down the first domino in its eventual starvation.' The team would all like to see regulations that prevent people putting locations up on wildlife social media. However, since many conservation groups insist on reported locations, they agree that this is never going to happen.

When I ask how they deal with people that just don't, or won't get it, Laura explains that they eventually report them to the council, or to the police. 'That's all we can do,' she says, adding that all dolphins within twelve miles of the shore are technically owned by the King. 'Maybe we should bring him in, too!'

Often, she tells me, ignorance is the excuse. 'I didn't know that I wasn't meant to splash them with my paddle, and all that. It doesn't work with me. I mean, if I went to Saudi Arabia, it's my responsibility to know whether it's OK to wear a bikini or not.' It's a good point.

The next time I push the packet of biscuits around, only I take one. I try to eat it furtively, an activity for which the crumbly digestive is not well suited.

The two of them then run through the menu of other threats: removal of food sources, run-off pollution from rivers, and micro-plastic ingestion. Joe reminds us that poisons accumulate incrementally the further up the food chain they go, and the bottlenose is at the top. Persistent organic pollutants that were banned back in the 1970s are still present in the sediment,[5] and therefore still make their way up the food chain, particularly with an animal that can live for over 50 years. Nitrates and phosphates washing down the rivers from farming activity and anything that is still leaking out of the old industrial mines, it all adds to the cocktail of ingredients out at sea that don't help.

And then, of course, there is that warming ocean.

'That's the huge elephant in the room all the time,' says Maddy. 'Climate change. We simply don't know how the food chain is going to react to the warming seas. Things maybe go north to find cooler waters, and they may get replaced by other species coming in from the south on the same basis.'

Laura points out that this could well include the great white shark. 'That might be great for tourism in the short term but could well sound a death knell for the bottlenose dolphin, who currently has no natural predators. Anyway, what happens when that northerly process stops, when there's nothing more coming from the south? No one knows. As it is, we're getting more algal blooms each year, as a kind of sign that things are changing, and not necessarily in a good direction.'

For the time being, the Cardigan Bay dolphins are doing pretty well. Everyone in this room is less than half my age, and I find myself drawn to asking these younger environmentalists how they deal with the burden of knowing what they know. I explain that we can't unsee what we have seen or unlearn what we have learned.

'It's never not been my life,' says Laura. 'I get a lot of my optimism from the advances in knowledge that people like us are contributing to, and from the expressions on the faces of the children who come

---

[5] Known as POPs, these are toxic chemicals that 'adversely affect human health and the environment around the world. Because they can be transported by wind and water, most POPs generated in one country can and do affect people and wildlife far from where they are used and released.' United States Environmental Protection Agency.

here. It's never not going to be worth fighting for.' She tells me that she still gets excited, even after fifteen full-on years, every time a dolphin appears in the view outside the office window.

Maddy agrees. 'It's the incredible benefits that what we do brings to people's wellbeing, our own included. I know that I am going to spend my whole career trying to make things better, and that alone keeps me hopeful.'

I offer the biscuit packet around one last time, again to no effect. Putting three digestives in my fleece pocket, just in case, I pointedly leave the remnants of the packet on the table, as if eating them would be the last thing on my mind.

A few minutes later, we are back at the sea wall, chatting to the volunteer standing on perpetual watch over the sea.

'Anything?' I ask.

'No. Nothing.' She is cold and damp but still cheerful. It turns out that I had chanced to strike dolphin gold in the early hours of the morning. They weren't really expecting any more today.

Laura introduces me to one of the beach-cleaning volunteers, a twelve-year-old boy called Adam, who is a recruit from the Stand for Nature Wales youth initiative and is equipped with a smart little picking grabber. She suggests that I ask him a gentle version of the hope question.

'I'm hopeful. Sure I am,' he says almost casually, after thinking about it for a second. He and I know that it is a stupid question. He has been picking cigarette butts and bits of plastic off the beach for months now, and he can see the value of what he is doing. He knows instinctively that just being involved gives him some tiny element of influence for his future. He knows that each friend who comes down and joins him on a beach clean will be another child who doesn't thoughtlessly chuck a sweet wrapper on to the sand. Being surrounded by people who care, he finds that he drifts gently away from those who don't.

'It's nice just seeing things getting better.'

As I talk to him, I find myself momentarily thinking a few months back to Bob, the old birdwatcher up at Ardnamurchan. Through their seemingly ordinary efforts, both are intimately involved in recruiting the next young person to stand up and be counted on behalf of nature, whether they know it or not. And it is the degree of success with which we recruit that 'next young person' and the next one, and the one after, that will define the future health of our biodiversity.

Laura tells him that he needs to get back to work, that the beach won't clean itself. He is having none of it.

'You promised we could play Top Trumps once we finished this bit,' he says seriously. 'And we all have to keep our promises, don't we?'

Ice Cream

South Wales

Magnified in the vast emptiness of the beach, the solitary ice-cream van shimmers out of the haze from at least half a mile away and I know immediately that I must stop there. Nothing has been so certain for me for weeks.

Although I can happily go months without buying an ice cream, it is as if the very presence of sand under my feet, gulls over my shoulder and a blue ocean beyond the horizon is unlocking an elemental need within me to stuff my face with freezing sugar and cream. And it is. Ice cream is an integral piece of the coastal jigsaw puzzle, mine at least.

This is the current end point of a line that stretches back over half a century to a holiday in Seaview on the north coast of the Isle of Wight, where ice creams were the vital architecture of my days. Looking back at it now, they were generally bribes, but who cares? Bribery has an honourable place in a boy's life. I can still see the twin pepper-pot funnels of the huge liner, the SS *France*, making her slow way out into the Solent from Southampton Water as I run down the early morning streets to buy my parents' friends' daily packet of Embassy Number 1 from the tobacconist on the front. No trade unionist could have protected the demarcation of his employment rights on this errand with more ferocity than I did, no employee could have been a more reliable timekeeper. I can still remember the daily feel of the tuppence reward money in the pocket of my red shorts and how, after five days, I would have enough to choose exactly the ice cream I wanted and at the time that I wanted it. Just the once, when the other children would be stuck with humble Woppas (4d), Sky Rays (6d) and Wafers (8d), I would buy my own precious Top Ten, with its nut, chocolate and vanilla. It mattered not the least to me that theirs were being bought for them, and that I had been offered one myself. All that counted was that I was free to choose the one I really wanted, the one that was out of the understood price range. For the first time, I had advanced myself by my own efforts, and a work ethic was born.

Holidays came and went, but the ice cream stayed. Nine years old became 63, the Top Ten became Mint Choc Chip, and 10d became £3.50, but no change really mattered; it was the freedom to choose that counted. So I buy the ice cream and walk on eastwards across the wide intertidal sands counting the jellyfish, lost in guiltless pleasure, and happily licking splashes of melting green before it drips from the cone onto my right hand and my shorts.

Only later in the day do I understand that this moment has come at a key time in my journey. It has reminded me that the pursuit of happiness is not a sin, and that the coast, especially the sandy beaches that plot this liminal line between the reality of dry land behind and the unknown sea beyond, offer millions of us each year the uncomplicated pleasure of a holiday on the sand. Sadly, much of what we nature writers occupy our time with is the business of recording decline. How precious, then, from time to time, to simply record that glorious Mint Choc Chip ice cream instead, for nature is not going to be restored by misery alone.

'That man is richest whose pleasures are the cheapest', as Thoreau is supposed to have said, and the beach remains a repository for simple joy, the sea beyond a bringer of simple gifts.

Right: The limitless beauty of the British coast.

Below: Sand sculptures on Sandwood Bay.

Feed pipes on a Sutherland salmon farm.

Sea eagle territory.

White-tailed eagle: back to stay? (Amanda Fergusson)

Puffin: more sand eels and fewer rats, please.

Native oysters in Loch Craignish. (seawilding.org)

The abundance that sea grass brings. (seawilding.org)

David Shackleton's 'plastic' garden.

Another Place: Crosby Beach.

Above: Spartina on Hoylake Beach.

Left: Bardsey Island: a 'thin' place.

Joe Wilkins and friend, Cardigan Bay. (Joe Wilkins)

Bottlenose dolphins off New Quay, Cardigan Bay. (Sarah Perry)

Left: Wyl Menmuir 'wrecking' on Hale Beach.

Below: The beauty of Cornwall: St Michael's Mount.

Above: Budd's Farm outfall. Better for gulls than swimmers. (Chris Pearsall Photography)

Right: The new face of activism: Rob Bailey (left) and Kevin Edwards.

Above: Leisure, in its many forms, is crucial in bringing back prosperity to the coast.

Left: Street art in Margate.

Above: The Thames Barrier. (Ivan Haigh)

Right: London's very own seaside.

Left: The Broomway: a footpath like no other.

Below: Abundance: pink-feet heading for the Norfolk stubble at dawn. (Sarah Farnsworth Photography)

Dawn and wind farms at Spurn Head. Gateway to the Humber.

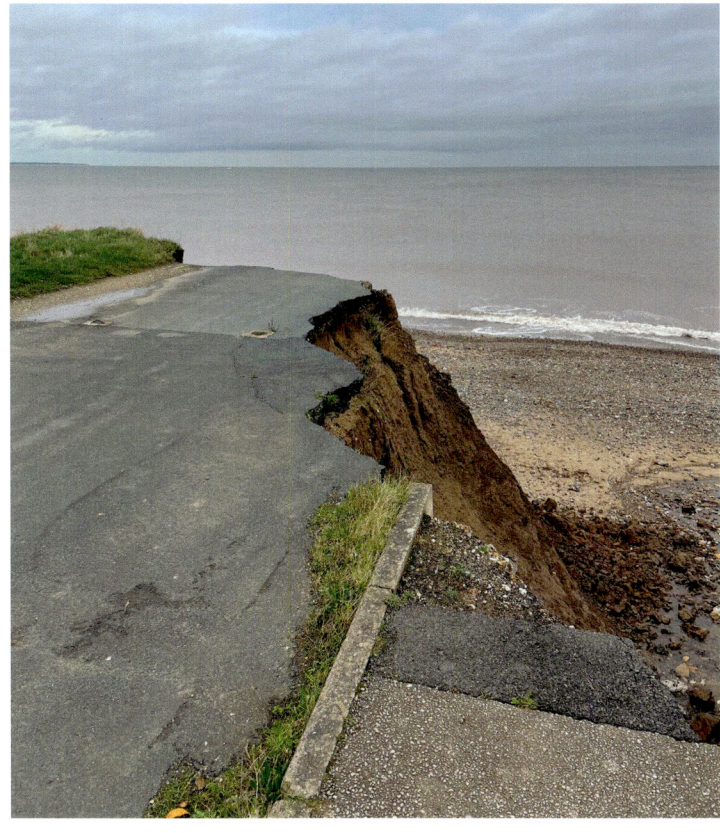

Holderness: the road went somewhere once.

Bempton Cliffs: last solid ground before Dover.

Gannet struck down by bird flu. Natural causes or a man-made disaster?

Above: The author on Holy Island. (Fiona Fell)

Right: View from Mark Newell's office: Isle of May.

Left: If boldness and determination could save a species: the wild Atlantic salmon. (Sarah Farnsworth Photography)

Below: The last man in Britain. (A photographer at Dunnet Head)

PART 3
EASTERLY

THE SEA AS BRINGER: THE SOUTH WEST COAST 7

October

> 'There's no such thing as "away". When we throw anything away, it must go somewhere.'
>
> Annie Leonard, Greenpeace USA

Sometimes, we can't even see what gifts the sea is bringing us.

Out there over my right shoulder, as I join the South West Coast Path at Minehead, is the biggest power station on earth.

At least it could be, and possibly should be. Second only in the world to Canada's Bay of Fundy, the Bristol Channel has huge tidal ranges of between twelve and fourteen metres, providing a viable source of potential power close to the centres of population where that power is most needed. Safe, barely visible, reliable and entirely predictable, I have spent quite a lot of my journey wondering why tidal power is still virtually non-existent in the United Kingdom. After all, it's not exactly as if we thrive within our current energy regime. So what has gone wrong?

'Essentially,' says Chris Binnie as we walk along the sea front at Minehead, 'it's never been cheap enough in relation to fossil fuel for long enough for us to adopt it. And then wind power came, and the same applied. The upfront costs are so huge that it needs government to guarantee projects, which is something they've not yet been prepared to do.' Chris should know. He's a fourth-generation water engineer, honorary professor at Exeter University and tireless campaigner for tidal power. He is leading a new bid to create a major tidal lagoon just off Minehead. It doesn't stand a chance of being operational for twenty or thirty years, but that doesn't deter him one bit. Nor should it. Harnessing power from the tide is a long game, stretching back to before medieval times.[1]

'There are three types of tidal power,' he explains. 'Tidal stream, like in the Pentland Firth, where fast ocean currents drive turbines, sea windmills, if you like. Then you have tidal barrage, as in La Rance in Britanny, where an estuary is dammed to block receding waters during ebbs; and then my own area, lagoons, where electricity is generated from the natural rise and fall of the tides within an artificially enclosed area.'

You could also mention harnessing power from the movement of the waves, by the way, but that would seem the most remote prospect of all. There have been various recent attempts at lagoons, from the Swansea Lagoon, where the UK government withdrew its support at the last minute on value for money grounds, to a new Severn Barrage that could generate 7 to 10 per cent of the UK's electricity, if only it can pacify the Port of Bristol about shipping flows and conservation groups about wader grounds.

There are always going to be tricky compromises. Back in the 1940s, when the great American environmentalist Aldo Leopold came up with his ethical principle that an action is right or wrong to the degree that it benefits or harms the well-being of the ecosystem, I suspect it was quite simple to work out which was which. These days, when we have screwed up so much of what was handed to us and have so little left, it is much more complex.

---

[1] A tidal mill is recorded in the Domesday Book, dating back to 1086.

For example, for the greater gain of helping mitigate man-made climate change and sea level rise by producing renewable energy, it may well be necessary to lose some wader breeding grounds or accept the death of a few seabirds in the unforgiving blades of offshore wind turbines. But that is not how public inquiries work. Besides, anyone trying to do anything that involves disturbance to the seabed these days will find themselves answering to a bewildering number of official and semi-official bodies who are trying to ensure that there is an acceptable equilibrium between engineering, economic, societal and environmental needs, and that their own specific interests are not subordinated to anyone else's. In our nature-illiterate society, these conversations are seldom nuanced, which goes some way to explaining why our infrastructure record is generally so grim in comparison to our European neighbours.

We are standing on the beach just east of the town, looking out at nothing, which is exactly why we are here. Chris tells me to use my imagination.

'Imagine a semi-circular D-shaped wall out there,' he says, waving an arm out at the murky channel, 'with a 14-kilometre diameter and a 22-kilometre perimeter. Now imagine five blocks each of 25 turbines, 125 in all, just poking up above the surface of the sea. The rise and fall of the lagoon inside will generate electricity on both the ebb and flood tides, on average for 21 hours a day, and should produce around 6.5 terawatt hours per year or, in layman's terms, enough to power two-thirds of the domestic needs of South West England. Day in, day out, every day. There will be two locks for boats to come in and out, and a new marina.'

There is something infectious about his enthusiasm. I try and fail to envisage how it might all look, so he helps me with an artist's impression that he has brought with him from the car.

'It's not just the power,' he continues. 'It means protection from flooding, storm surges and erosion; it means employment in both construction and operation, probably a steady hundred local jobs once it is up and running; it means increased tourism through recreation and, very possibly, a new marina at Minehead.'

'What about the fish and other wildlife?' I ask. Fish are notoriously inconvenienced by dams, and this looks mighty like a dam.

'There's obviously mitigation to be done, as there is in any project, but the site has been chosen to avoid spawning rivers and wader breeding grounds. Nothing is perfect,' he adds. 'and we need to do more modelling on things like water quality, eel migration and sedimentation. But overall, we believe that it's the best of the eight other sites that are being scoped around the UK at the moment.'

I mention the cost, that trumpeting elephant in the room in most previous projects.

'£8.5 billion,' he says. 'and six years to full power. These things are usually compared using costs per megawatt hour generated, where we currently are not fully competitive with current offshore wind bids, but are very similar to the new nuclear power station at Hinkley Point C over there.' He points westward to a blur lost in the sea mist a few miles east. 'But they have a life of 50 or 60 years against our 100. And we don't produce dangerous waste. The problem is not the cost, per se, but the way the private sector use rates of return with discounted cash flow that don't go beyond 30 years. Meaning it needs government support. Meaning it needs government vision. The debate is moving in our direction. Eventually, it has to.[2] Apart from anything, as we slowly move to gas as a leading automotive fuel, ultra-reliable tidal power would be perfect for creating hydrogen.'

I'm no scientist, but I find that I am with him, convinced. I love that, in a country that traditionally likes small projects and fast returns, someone is thinking really big and really long-term. I love that maybe one day, someone really will harness the predictable

---

[2] A year later, with the formation of a commission to look at options by the Western Gateway Partnership, it has started to look as if this is happening.

twice-daily movement of that vast wall of moving water I have been following all the way from Cape Wrath. I love that, maybe one day, we won't be so reliant on burning subsidised fossil fuels to power our hungry lives. But most of all I love that the man whose idea it principally is, the man patiently explaining to me how it will all work, is 22 years older than me.

At the age of 85, Chris Binnie is breathing daily life into a vast project that has no hope of being operational for 20 or 30 years. I tell him that he has the mindset of a medieval cathedral builder, and he smiles, chucking a pebble into the sea for his dog to chase.

'It will come,' he says.

The soundtrack of my journey, when I am on foot, is the white noise of rolling waves and breaking surf in my right ear. Always my right ear, as I walk anticlockwise, like a sheep walking endlessly the same way around a steep hill. My relationship with the sea has changed over the last few months, and it will change again. For now, it is the constantly present, constantly moving amniotic fluid that confirms me in my islander status, the only 'always' of my journey. Depending on its mood, its colour palette varies from Caribbean aquamarine all the way through to coal black, sometimes within the space of just a few minutes, but I love it most when the sunlight just before rain slants in brilliantly to illuminate a lone, laundry-white gull against the charcoal-dark clouds. Depending on where I am, its character varies from the barely conspicuous lapper of the shoreline in some southern salt marsh to the voracious consumer of cliffs around Bridlington or Bridport. Because I am largely alone, the sea is my restless companion in a way that the coast itself never can be. It is the vast heat-exchange machine that connects me and my coastline with every other coastline in the world, over a tiny fraction of whose limits I am now travelling, and with every other person on those far-off shores.

I have been walking on the South West Coast Path on and off for over four decades, so this time I concentrate on the sections I haven't walked before, in the process reminding myself that 15 miles on these rises and falls are worth 20 or 25 in most other long-distance paths. The scrum of summer tourists is long gone by now, leaving the midday world to retired people with their dogs and young surfers with their boards. I dip in and out. I look up to the sky and out to sea, but rarely inwards towards the land.

Mine is now a world of mud and hillside steps and retired tin mines, of rain and narrow steep paths through the gorse, and of stonechats flitting around in the margins, always perched high, always in pairs; a world of wheatears, wind-hovering kestrels and wind-bent hawthorn bushes, as old as history, and of crows and cattle. At Portreath, I meet a New Zealander who is working his way around the world's great long-distance paths.

'And is this one of them?' I ask.

'I guess so,' he says, without commitment, and then adds: 'At least I'm not going to die on this one!' It turns out that his adventure has provided more in the way of peril for him than he may have bargained for. We swap sympathies for our respective countries' lack of success in both the recent cricket and rugby world cups, and then arrange to meet at a pub that evening, where we discuss red cards, middle age and Australians.

Out of season and after dark, the picturesque fishing villages of Cornwall seem devoid of light and life for the most part, to each cottage door its tell-tale key box,[3] to each upstairs window its darkness. This is a buy-to-let world these days, where cottages with names like Neptune House, Net Loft and Gull's Way prepare themselves for a silent winter, chairs stacked neatly on kitchen tables while the sea thunders remorselessly against the harbour walls in the darkness beyond. The locals who work here, by and

---

[3] The diagnostic marking of a seaside AirBNB establishment.

large, move inland for the night. After all, the average price of a house in less fashionable, inland Camborne is £256,000, exactly half of one in St Ives.[4] Second homers don't send children to local primary schools, so you can guess the rest.

One of the added human problems of the coastal communities is that most of the homes in the beautiful bits tend to be owned by people who don't even want to live there. At least not themselves, and not full time. Helpful figures on someone's domestic balance sheet they may be, but contributors to social cohesion they most definitely are not.

And still the sea brings in its treasures.

On Towans Beach at Hayle, I have arranged to spend an afternoon beachcombing, or 'wrecking', with fellow writer Wyl Menmuir, and we start with a peanut in its shell.[5] I love that we start with a peanut. Why wouldn't I? My entire journey started with a peanut.

'It's almost certainly come across the Atlantic from America,' he explains. 'On the Great Ocean Conveyor.' The prospect of this lengthy odyssey thrills me more than I expect, and I put it for keeps in a side pocket of my backpack, along with the out-of-place feather of a great-spotted woodpecker. Wyl has done this for years, ever since he migrated to Cornwall from his native Stockport a decade ago. He points out a fisherman's kiss, the local name for a small offcut from the process of mending fishing nets that then gets lost, blown or washed over the side of fishing boats and quays. It is one of the commonest things found on the beach and, if not recovered, becomes a leading ingredient of microplastics that are ingested by

---

[4] Rightmove, 2022 prices.
[5] His beautiful book *The Draw of the Sea* is a good place to start on your own personal sea journey.

a high proportion of the seabirds and fish off the Cornish coast.[6] All afternoon, we find more.[7]

Sometimes, as a species, we listen but we don't hear; we see but we don't absorb. There is now a staggering weight of plastic continually making its presence felt on the British coastline, our national share of the eight million pieces that make their way into the global ocean on a daily basis, and the twelve million tons that are dumped in it each year.[8] Plastic has been as constant a companion to me on this journey as the gulls and the surf, strung in colourful ribbons along the tideline and bobbing in the rock pools no matter how isolated the shoreline. That's just the stuff we can see, the stuff we therefore mind about. We leave largely unremarked the unseen microplastics from our car tyres, our laundry, our personal care products and from the rest of our daily lives, and which go on to get ingested by fish, mammals and seabirds; particularly seabirds, over 90 per cent of which are now reckoned to have pieces of plastic in their guts.[9] With some of the cleanest and most widely available tap water on earth, the British still manage to get through billions of plastic water bottles each year. Wyl is acutely aware of these statistics, as am I. We all are. And yet a combination of laziness and short-term convenience ensures that the problem increases every year.

We find a cooked lobster claw ('posh beach barbecue'), a coil of blue rope covered in goose barnacles, and then a tiny false jellyfish, similar to the Portuguese man o'war, called by-the-wind sailor,

---

[6] The creator of an art installation in 2018 managed to recover over 10,000 fishermen's kisses from just one beach in Cornwall.

[7] Next time you are at your optician, if you have one, remind them that Newquay company Waterhaul creates frames for glasses and sunglasses made from recycling ghost-plastic from the fishing industry.

[8] Surfers Against Sewage (SAS).

[9] *Threat of Plastic pollution to seabirds is global, pervasive and increasing.* Wilcox et al. Paper for PNAS, August 2015.

faded blue on the sand, complete with the sail that brought it here. It is beautiful, and I find myself strangely moved that it has arrived here, as have all the inanimate things, utterly at the behest of the winds out at sea.

Wreckers have a particular look about them, and I can quickly tell that we are not the only ones up to it on the beach this afternoon. Slow walk, hunched shoulders, downward-facing head and the occasional stoop and discard are the signs. 'The joy is in the searching and not the finding,' Wyl explains, 'At least for me it is, even though there is always that tiny chance that something extraordinary will turn up.' He has a point. Just 30 miles up the coast that very morning, a 54-foot long juvenile fin whale had washed up dead on Fistral Beach near Newquay, although that may have been beyond the capacity of even the most committed wrecker.

Wyl talks about how he finds it the most democratic of activities in a place where so much activity, surfing for example, has been subsumed into brand-building and costly marketing. He carries a large and robust backpack into which he deposits the plastic rubbish that he finds, and admits to a tiny, karmic expectation that this might lead him to being rewarded by finding something extraordinary himself. A sea bean, which is a tropical seed adapted for travel on the ocean currents, is right up there in the hierarchy of desirable finds; a sea bean that has made its slow way up the prevailing current from the Caribbean Sea, maybe taking two years over the journey, still intact, still theoretically viable, until it comes to rest on some eastern Atlantic beach.[10]

Around us the beach life rolls on: a surfing class gathers in a semi-circle around their teacher; a couple walk their two mad spaniels at the tideline. Two pied wagtails run ahead of us for

---

[10] To learn more on the fascinating world of drift, you might read Sally Huband's *Sea Bean*. Apart from anything, you will learn that a bottle with a message in it can cover the same distance in a little over a year, as enough of it is above the water to be blown on by the wind.

whatever our progress might disturb. We walk on. As if to caution us against any over-optimism at the slow recession of the effects of avian influenza, we find three dead guillemots, equally spaced towards the north end of the beach like stubby, elegant brown and white jars. Spent and faded shotgun cartridges ('American, again'), a length of blue farming hose, some dogfish egg cases and a little yellow sandcastle mould. 'I just want us to stop selling crap,' he says. 'Little beach spades that are so flimsy they almost count as single-use plastic. Surely we have advanced further than that these days? I mean, we know exactly what plastic does to the wildlife.'

Apparently not. He picks up something that could be a Vaseline jar or a chewing gum container, then a balsawood knife, then a shaving-gel can and the lid of a water bottle. The fascination, he says, lies in the uncertainty and the idea that the mysterious sea, given two identical things set afloat in the same place at the same time, could eventually present them in different continents and separate centuries, or never present them at all.

An incongruously smart woman clutching an artist's folder approaches us, and we assume that she has been painting a landscape.

'Not at all,' she says. 'I'm a calligrapher. You know, wedding place-cards, golf club Christmas dinners, that kind of stuff. But it's not so busy at this time of year, so I come down here every day looking, just in case. I've just met a lady who's found some Lego.'

I miss the vital wrecker's cue, but Wyl doesn't. To say the word 'Lego' to a Cornish wrecker is akin to whispering 'gold' to a forty-niner.

'Lego?' he asks, looking as disinterested as he dares. 'Whereabouts?'

The local fascination with Lego dates back to 13 February 1997, when a series of freak waves struck the New York-bound cargo ship *Tokio Express* twenty miles off Land's End and dislodged 62 containers into the sea, one of which happened to be packed with mainly sea-themed Lego pieces, which have been turning up on

Cornish beaches ever since. Many more concerning contents were lost, such as an irritant superglue and various hazardous chemicals, but it was the Lego container that has been the gift that has kept giving ever since. So much so that it has even become the subject of a BBC documentary and part of an academic study of ocean currents.[11] American oceanographer Curtis Ebbesmeyer has tracked the Lego ever since it went overboard, and points out that some of the pieces have had time to drift over 60,000 miles, meaning to any beach on earth, even though it has mainly remained a Cornwall phenomenon. Every now and then, someone from Florida shouts out a claim, but to date it has generally come out of a set that was not even in the container.[12]

Science determined that most pieces were of a formula of polymers that caused them to sink straight to the bottom, where they would continue to move around in the currents and even become the substrate for someone else's habitat. But many pieces floated. Presumably aware of the publicity potential, Lego released the manifest of the 4,756,940 pieces that were in the container – 353,000 daisies, 28,700 lifeboats, 52,000 propellers, 50,000 sharks, spearguns, scuba tanks and seagrass – an act that ended up creating an informal hierarchy of what was most desirable to find, culminating in the black octopus, the wingless green dragon and the shark. To date, not one of the sharks has ever been found.[13] The flotsam Lego pieces had become part of an art installation at the Turner

---

[11] Radio 4: *Lego Overboard*.

[12] Lego is not alone. In 1991, no fewer than five containers of Nike trainers broke free from the *Hansa Carrier* somewhere between Seoul and Seattle, releasing 61,280 shoes to their oceanic wanderings. The following year it was plastic bath toy sets. And that's before all those yellow plastic ducks.

[13] Stop Press. Devon fisherman Richard West scooped one out of his nets twenty miles south of Penzance in August 2024. Finally, nearly three decades after the container was lost, a Lego shark had been landed.

Contemporary in Margate by the time I got to Kent. Somehow, it was so much better in the imagination.

Wyl and the calligrapher briefly exchange their Lego notes.

'If you find that black octopus,' she says, 'you will scream, won't you? I mean really loudly!'

Plastic lasts for ever, and much of what does not end up round a turtle's neck or ingested by a shearwater is eventually returned by the tides and currents to the species that created it.

The scale of the problem is immense. In 2017, the Marine Conservation Society found 718 pieces of litter, the vast majority plastic, for every 100-metre stretch of coastline in the survey, each piece a tiny share of the 14 million tons of the material that is dumped in the sea each year, making up 80 per cent of all marine debris.[14] The day is fast approaching when the weight of plastic in the oceans may start to challenge the weight of fish.

Two days after my wrecking adventure, and as far from Cape Wrath as I will ever be on this trip, I have arrived in the first town in Britain to have been granted 'plastic-free' status. I am a bit sceptical at first, as my experience has been that most of these schemes are driven by a small group of inspired activists and widely ignored by the rest. Mainly businesses, and especially supermarkets. As American novelist Upton Sinclair put it long ago: 'it is difficult to get a man to understand something when his salary depends on his not understanding it.'

But not here. Not in Penzance.

'I suppose a couple of things came together to start and then enable it,' says Rachel Yates, the woman who has been driving the project from the start, who is sitting opposite me in a café in the town. 'It was 2017. First, I was working with Surfers Against

---

[14] International Union for Conservation of Nature.

Sewage, who wanted to run a pilot to see if they could dramatically reduce single-use plastic usage at source in particular places. Secondly, I lived in Penzance and knew the community well. And third, it was at the exact time that David Attenborough's *Blue Planet* was being shown, so the interest was already out there.' The power of well-made nature documentaries to be the seed bed of activism is often underestimated.

Single-use plastic is a compelling target. Unbelievably, 36 million plastic bottles are used in the UK each day,[15] only half of which get recycled; two and a half billion coffee cups are bought and thrown away each year, of which just 1 in 400 get recycled. Half a million of them simply end up on the streets every day.[16] The sale of single-use shopping bags has reduced by 75 per cent since the 2015 legislation, but that still leaves half a billion. Once you are talking about things in billions, you are probably entitled to believe that the country is drowning in the stuff. And once you are drowning in the stuff, you are entitled to believe we are mad to go on producing it for inappropriate use. For a country that survived happily for decades, centuries even, without them, and would do so again tomorrow if it were given the chance, the 7.7 billion plastic bottles used in Britain each and every year comes as the greatest proof of all that some businesses consistently put profit ahead of planet, and therefore why we all need to be activists.[17]

Back in Penzance, what was surprising to everyone was the speed of the uptake, something Rachel puts down to it being the right plan at the right time. 'The absolute key was to be positive, and not to lecture. It's a civic pride thing. We first targeted all the

---

[15] Recycle Now website.
[16] House of Commons Report: Disposable Packaging Coffee Cups. 2018
[17] Water UK. Another source puts it at 13 billion. The world total is somewhere around 482 billion, according to habitsofwaste.org

people we needed onside to give it credibility: the town council, who have been supportive from day one, local businesses and then all the potential community allies like schools, libraries, churches and sports organisations. And we targeted the worst items, not all of them. We kept it as simple as possible: just getting rid of single-use plastics. We ran events, gave advice and created mentors from the people who were already engaged. Within three months, we had our accreditation from SAS. Six years later, there are about 200 towns with the accreditation, and another 926 or so committed to it.'

I ask her what it all actually means, what it looks like on the ground.

'Various things,' she says, ticking them off on her fingers. 'Organised beach cleans. One-off cleans after a storm. Removal of packaging at retail. Persuading cafes to swap single-use coffee cups with reusable ones. Getting shops to persuade their suppliers to abandon plastic in their packaging.[18] Even getting accommodation providers to provide Tupperware containers for tourists to use on their picnics. We got the local festival to replace single-use cups with a giant water bowser from which people could refill their own bottles.'[19] She warms to the task. 'Three years in, and with a previous average of 30,000 cups per festival, that alone has saved 100,000 cups. Some of it is as simple as demonstrating how much money people will save if they do things in a slightly different way. We even have dressmakers running old-fashioned repair services, and thrift and vintage shops have blossomed in the town.'

I mention the big Sainsbury's out there on the eastern roundabout.

---

[18] The town's cookshop, Hobbs, received a prestigious industry award in 2018 for its constant battle to persuade its suppliers to reduce plastic packaging. In another life, I was one of their suppliers, which made it doubly satisfying.

[19] Golowan: Penzance's 'Festival of Midsummer'.

'It's difficult,' she admits. 'The staff are all for it and really supportive, but it's progressively tougher for them to get real change made further up the supply chain. So we took to going in with a huge bin, and asking customers on their way out to discard the packaging that they would rather not have to deal with and didn't really need in the first place. The store was completely behind it, by the way. And by the end of the morning, people could see for themselves the huge amount of unnecessary single-use plastic. I think the penny eventually drops, and they are certainly using less plastic on their fruit and veg these days.'

The biggest reverse she has faced was in the aftermath of Covid, where many retailers found the health aspect of wrapping things in plastic beguiling. She reminds me that deprivation is high in Penzance, and that the cost of living crisis is real for everyone, inevitably pushing campaigns like hers further away from the oxygen of publicity.

As sensitively as I can, I mention the fishing industry. This, after all, is a tight community into whose DNA fisheries are woven. All through my journey south from Cape Wrath, I have walked along beaches full of their ghost plastic, and I can't help wondering where they fit into a project like this. Rachel says that they are supportive, to the point of bringing in crates full of plastic that they have brought up in their own nets.

'At Newlyn, next door,' she adds, 'they even make the town Christmas tree out of ghost fishing material each year, and they decorate it with things like scallop shells.'

Penzance is just one town in a country that is groaning under the weight of its discarded plastic, but I head eastwards, eyes wide open for those black Lego octopuses, all scepticism assuaged. As all of us in this environmental business must constantly remind ourselves, in order to avoid madness, the journey of a thousand miles starts with a single step.

People like Rachel can only walk alone so far.

The Lifeboatman

Lyme Regis

Cornwall gives way to Devon, and Devon to Dorset. Rocks give way to sand, and sand to crumbling grassy cliffs. Always movement, always change.

Two things start to dawn on me as I move east into Lyme Bay. First, how strange it is that we call this long and varied coast, with all its indentations, blurred outlines, movement and porosity, a 'line' at all; there is no line about it. It is one of the least linear things that I have ever seen. An octopus has more lines on it than the British coast. Secondly, I come to realise that what just about every town I have visited or just walked through has in common with the next one, and the last one, is a Royal National Lifeboat Institution (RNLI) station. Down the west coast from Lochinver to the Lizard, along the south coast from Salcombe to Selsey, and back up the east coast from Sheringham to Stromness, there are 238 British RNLI stations in all, involving nearly 10,000 volunteers who, between them, get called out (the technical term is 'a shout') an average of 22 times a day, and save altogether about 500 lives annually in the process. So much of the coastal life that I have been observing over the months is enabled largely because of the safety blanket provided by the unpaid men and women prepared to risk their lives to rescue people and their craft when things go wrong. And the risk is real. Anyone in Penlee can remind you of that. On 19 December 1981, the coaster *Union Star* floundered on its maiden voyage from Holland to Ireland. In hurricane conditions, the Penlee lifeboat was deployed to assist. Having rescued four passengers, the crew went back for more, and were probably crushed under the much larger coaster. All eight crew were from the village, and all eight were lost, leaving twelve children under ten years old fatherless.

Lyme Regis is the twenty-sixth RNLI station I have passed on my route (there will be at least 30 more before I am done), but the first where I have arranged to meet a volunteer. To tell any story of our coast without weaving the lifeboatmen and women into it would be more insult than oversight.

Before he changed roles, Nick Marks was the man who decided whether or not the Lyme Regis lifeboat launched once a shout had gone out, 'supposedly the calm head that could judge better than the adrenaline-filled volunteers whether or not the balance between the coming task and the risk was a proper one'. These days, he is the deputy communications officer, still an important cog in a giant wheel that needs to raise around £175 million a year to keep providing its service. Well under 1 per cent of this comes from the government, but it seems that we are still an island nation at heart, happy for now to fund it ourselves without being forced to.

'Did you ever say no?'

'Just twice. Happily rightly in both cases.' He smiles as he recalls his relief at getting the news that the previously unreported 'body' that had been spotted by a yachtsman fifteen miles to the south-east turned out to be, as he thought it would, no more than a large cluster of seaweed.

'Are the volunteers hard to manage?' In my experience, the type of person who might volunteer to risk their life on a regular basis to help someone in trouble would be strong on independence, and possibly short on being good at taking orders.

'Initially, yes,' he says, 'but that was me, not them. I had been a Royal Naval officer, where everything I asked a subordinate to do was backed up by the 1957 Naval Discipline Act. I discovered quickly that it doesn't work like that in civvy street, and it sure as hell doesn't in a volunteer organisation!' Having been a soldier once myself, I know only too well what he means.

We look at the notice board, with its Code of Conduct and its RNLI Volunteer Commitments posters, and its charts of who has achieved what level of training competence. There are as many as 39 people on the books, split among seagoing, shore-going and trainees. He runs through the names, humanising them by listing what they do in their other lives.

'Groundworker, musician, painter and decorator, architect, paramedic, sales consultant, butcher, chief executive. He runs a water sports company, he's primary school teacher, she's an artist. Most new volunteers, 90 per cent, I think, have no sea experience whatsoever. Among new recruits, it is generally quite easy to spot the ones who will make it through the two-year training programme: there's a sort of powerful curiosity and energy about them, and a reluctance to leave at the end of a session.' It is, as it has always been, a community within a community, vigorously local within a solid national framework.

From downstairs, the raucous sound of one of the two 115-horsepower engines of the station's lifeboat going up and down through its servicing routines means we have to raise our voices to be heard. In a warm and dry training room with a mug of strong tea in my hand, the noise is a tiny and welcome connection to that other world of call-outs.

'Has it got harder? You know, with health and safety, more people out there doing more challenging things, house prices, etcetera.'

'I guess the answer to all of those is a qualified "yes". Most of health and safety I get. We have an absolute responsibility to keep our people alive and well, which means managing risk in a hostile environment. It actually impacts more on our fundraising activities than what we do at sea. But they have stopped us recruiting under the age of eighteen, and that has removed a really reliable reservoir of future talent that we have always previously had, and benefited from. House prices are another thing altogether. Because volunteers have to be able to get to the station within ten minutes of a shout, they have to live in the town. House prices in the town are high and getting higher, meaning that it is highly unlikely that anyone in their twenties and thirties will be able to be on the books. These days, most of our recruits are well into their forties.' In other words, the poorer the town, the younger the crew.

Through the window, and out beyond the famous Cobb, I watch the calm grey sea under the lowering sky, the pastel shades of beach huts stretching eastwards and the rising and falling of the ever-present herring gulls. I tell him that I find it hard to imagine the brutal storm conditions that they must have to face.

'Actually, many of our call-outs – we've had about 50 this year – are on calm, sunny days and have nothing to do with the deep sea. Walkers cut off by the tide or landslips, people falling on rocks, paddle boarders getting swept away. Those are the ones on the increase.'

We go downstairs to look at the boat and the crew room, each space within it stacked neatly with the inner and outer clothing that a volunteer will grab and put on for a 'shout'. 'It comes to about £1,800 of kit per person,' he says, and then adds: 'Just check the weight of this life jacket. Used in anger, it will support the volunteer and three extra people.' I do so, and it is almost exactly the weight of my pack.

I could ask questions for hours more, but he has a job to do, and I have far to go in what remains of the afternoon; rather too far, as it later turns out. So instead, I ask him how they cope when things go wrong, which we both know is a euphemism for someone's death.

'We manage,' he says deliberately. 'We have a very good system of trauma counselling but, as you will know from being a soldier, there is no therapy as good as being able to talk about it with the people who were with you at the time. It's something that the RNLI is very good at these days.'

'Is there a worst moment? For you?' I ask, adding that he absolutely didn't have to answer that question if he didn't want to. To my surprise he leans forward, and almost imperceptibly, he starts to well up.

'The worst is when you can do nothing. And when it happens right on your doorstep as it did one day in 2011, when a severely disabled twenty-year-old girl, strapped into her wheelchair, must have accidentally caught the joystick on her chair with her arm, and went over the side of the Cobb into a dozen feet of water. The boatmen on the scene were already diving down to help her when the first of our volunteers ran along the Cobb, not even waiting for the boat. But it was all in vain. The wheelchair was too heavy to lift in time, and she was too securely strapped in to release, and she …'

He doesn't finish the sentence. He doesn't need to. Later on, I read that it took her parents four years to come back to the town from their native West Midlands, but come back they did, and they did so to say thank you and to help raise money by selling yellow plastic ducks for the town's Lifeboat Week. I sense that it is a treasured two-way connection.

As I am putting on my pack to go, he remembers that this conversation is for a book. 'Can I ask for you to include a safety message, please, even if it looks out of place?'

'Sure,' I reply, wondering how on earth it could be out of place.

'Two things,' he says. 'Be wary of the coast itself, not just the sea: erosion, rockfalls, the changing shape of beaches, that kind of thing. And can you suggest that everyone looks at our "float to live" video.[20] It's a really simple drill that could save your life if you suddenly find yourself gasping for breath in cold water.'

I write it down and head off into the drizzle, past the people in blankets in their deckchairs, the foreign students taking selfies and on into the long arc of the Jurassic coastline ahead.[21]

---

[20] Visit https://rnli.org/safety/float. It will take about two minutes of your life, but might end up saving it.

[21] The RNLI relies predominantly on charitable funding. During my coastal journey I had many stories of people whose lives had been saved by the work they do, and I was just one person among thousands walking these shores. If this piece has inspired you just a tiny bit, please consider supporting them yourself. You can do this by accessing the following link: https://rnli.org/support-us/give-money/donate

A BAY OF SEWAGE AND SEX-CHANGING FISH

8

The Hampshire and Sussex Coast: October

> 'There is sufficiency in the world for man's need, but not for man's greed.'
>
> Mahatma Gandhi

My eastward walk along the coast from Southampton is something of a hundred-mile counterpoint to an earlier westward walk my wife and I once did along the South Downs Way from Eastbourne to Winchester, also a hundred miles. Back then and up there, I remember the light airiness and the ever-present sense of history of an old hilltop path that people had trodden for millennia but where we were more or less always on our own, aside from the skylarks; right now and down here, it is all cars, trucks and buildings, and the underlying hum of busy humans going about their lives. At the other end of the island from Cape Wrath, I have reached the concrete heart of our coast.

But to walk the stretch of coast between Portsmouth and Brighton is to find shingle, too, and sea kale and stripey beach huts; it is to pass tamarisk trees, oystercatchers and assertive, hardy nudists in the spiky depths of the dunes; it is to see groundsel emerging from the cracks in the pavement and dusty *rosa rugosa* thriving on the edges of the shoreline car parks. Walking ever eastwards, it is to hear the thrum of engines in the grubby old fishing boats and the clanking halyards of a thousand gleaming yachts in a score of marinas and boatyards, and to watch serene paddle boarders in the evening sun and the brave kite-surfers flying over grey waves in the half-light of a windy dawn. Near Littlehampton, there are elderly tartan-clad golfers out on the links, and young girls dancing to a huge radio while fishing for crabs off the wall of the town's harbour; I see egrets where once there were black-headed gulls, and executive villas where once there was a salt marsh; there is the glittering light of an artist's seashore, and the towering thunderclouds commuting in from France.

This manages the trick of being at once a busy coastline and yet seemingly in retirement, where much of the money and opportunity has drifted inland with the river tides. The dawn of cheap holidays abroad and a once world-class transport infrastructure that failed to survive the Beeching railway line cuts of the early 1960s, brought in the years of decline, and it went downhill more or less from there. Pressure on already stretched local services has been exacerbated by the high proportion of people choosing to live out their old age by the sea, coupled with the low and unreliable wages that come with the care and hospitality sectors that result. It is widely acknowledged that, sooner than we would like to think, the increasing number of old people in many seaside towns will coincide with the decreasing number of carers to look after them, with potentially catastrophic consequences. If you look at an epidemiological map of the prevalence of coronary heart disease in Britain, itself a pretty good indicator of local health outcomes, there is a bright red ring around the entire coastline.[1] You would find the same with smoking, alcohol use, mental health and self-harm.

---

[1] Page 8 of Chief Medical Officer's Annual Report 2021 'Health in Coastal Communities'

The rise of coastal poverty seems to be in lockstep with the decline of the once iconic seaside pier, where there are now almost more that have been lost than survive.[2] A House of Lords 2019 report ('The Future of Seaside Towns') unambiguously concluded that resorts 'have felt isolated, unsupported and left behind' and that 'the British seaside has been perceived as a sort of national embarrassment'.[3] It doesn't help that, as one ice-cream seller put it to me, '50 per cent of my potential customers are fish'. Ironically, it is the recent discovery that working from home is a sustainable practice that may be starting to breathe new life into these places; not that surprising when you compare house prices and air quality in somewhere like Margate or Middleton-on-Sea to, say, those in London.[4]

And yet ... in Bognor Regis, there is famously deprivation, but there is also a rakish air of something bad having been recently survived, a sticking of two resigned fingers at the doom-mongers and earnest social commentators. They were writing the town off half a lifetime ago, when I earned a winter's living as a teenager in the brussels sprout fields to the north, and they are writing it off still. The shop fronts reflect just how little ever stays the same either with people or what those people want: the Sofia Market, the Turkish barber, the Vape store, phone shop, Stonepillow, Sports Direct, Costa and Greggs. Bognor is at once a powerful and fragile reminder to this coastal traveller of the moving nature of Englishness, where identity is a product of perspective and not

---

[2] British Piers Society (who knew that we had one?) website lists sixteen currently surviving, twelve lost and none currently in development in this section of the coast.

[3] A good place to learn more about the causes and effects of the decline of the British seaside town is Madeleine Bunting's book *The Seaside: England's Love Affair*.

[4] You can also read much more about coastal problems and their possible solutions by visiting the website www.coastalcommunities.co.uk

simply birth. Same in Worthing, which, for all its reputation as 'God's waiting room', is full of newness and building; same with Brighton, a city of apparent hedonism that these days takes itself extremely seriously. That's the thing about our coast: no place ever lives up to, or down to, its convenient stereotypes.

I spend most of one lazy afternoon, on what turns out to be one of the last fine days of the autumn, sitting on a shingle beach near Pagham, looking out at the Channel and reading a book about the very pebbles I had been ploughing my arduous way through, and am now sitting on;[5] learning that they have arrived on this beach not as one simple consequence of local geology and waves, but variously from a mixture of local erosion, longshore drift, faraway river catchments, tide and even ice that disappeared 10,000 years ago. That's why they come in so many different shapes, sizes and colours. That's why the whole coastline is this dynamic, constantly changing character. Maybe nothing on my long journey highlights to me my own transience on this island of mine more than the knowledge that each solid, shiny pebble under my feet was 'born' somewhere, is living and will eventually die just as I will. It is just that the stone doesn't breathe, and the timescale is far, far longer. That shingle beach carries in its stones and fossils the collective memory of a thousand influences on an island that has itself moved 8,000 miles to be here. When you understand that, you understand that this Anthropocene era that we are putting the planet through will, like us, be of little or no consequence a million years from now. It will just be memories on the outer skin of the rolling blue marble.

---

[5] *The Pebbles on the Beach*, Clarence Ellis, 1954. A minor classic, if you have the sort of enquiring mind that can't quite believe things like stones are as simple as they at first look. When Ellis wrote it, most seaside towns had a lapidary that you could take your stones to for cutting and polishing.

In an echo of my time at Ardfern, and on the Isle of Arran all those months ago, I'm looking out at things I cannot see, this time from the roof of a low building on the south-western edge of Langstone Harbour, on the edge of Portsmouth.

The gorgeous view in front of me is illusory: beneath the glittering sea out there, and the crews of the colourful sailing boats enjoying it, lies as wrecked a marine ecosystem as you could find anywhere in Europe. And not just wrecked, but continually contaminated with an annual average of 67,000 minutes of storm water discharge (aka sewage) from the local water company.[6]

Eric Harris-Scott is a young marine biologist, and is the Solent Project Officer for the ocean charity Blue Marine. It is fair to say that what he doesn't know about native oysters isn't there to know. If you're a native oyster, you probably want him looking after your interests. Undergraduate, Master's student, project lead: native oysters are all he has ever professionally done, which works rather well, as his job is reintroducing them to the seabed at scale. He is trying to reverse an extraordinary decline which seems to have started late in the Industrial Revolution and continued apace ever since; in 1850 at least 500 million native oysters were sold in London alone each year and, far from the luxury fare of the rich that they are today, were very much regarded as a staple for the poor.[7] Archaeological digs from well inland have shown that oysters have been an important part of our diet for over 2,000 years and, as recently as 1970, the industry they supported sustained over 500 jobs on the south coast alone. And then there's all the water filtration they were doing. Restoring native oysters is not just some environmental obsession.

These days, native oysters could act as a poster child for everything that can go wrong with a species: pollution, disturbance,

---

[6] Friends of Langstone Harbour newsletter, September 2023.
[7] *Rewilding the Sea*, Charles Clover, Witness Books, 2022.

destruction, development, disease and, not least, the flourishing of its quicker-growing (and therefore more commercially attractive) Pacific version. The case for their restoration is a strong one, not least of which is the fact that they are a keystone species, in that many other sea creatures make their home on an oyster reef,[8] something I'd seen firsthand at Ardfern. Add to that the filtration they do and the seabed stabilisation they carry out, and the native oyster makes the strongest case for immediate restoration. There is an occasional temptation to believe that this kind of project is a bit of a vanity exercise for an attractive single species, but nothing could be further from the case – this is about restoring an entire ecosystem.

'Look out there,' he says, waving an arm northwards at the sunshine on Langstone Harbour and the hills beyond. 'We've got a quarter of a hectare of native oysters out there, on artificial reefs that we have had to build on to the seabed, and we have plans to make that up to four. This part of the harbour is 28 hectares, so there's lots of room. Then add to that the oyster reefs we have recently laid in the River Hamble, and the ones we are planning to lay in Chichester Harbour over the coming years, and you begin to see that we are really thinking about this at scale.' I understand what he means: a lack of scale is always the limiting factor on conservation work, and I particularly love that the seed money for this venture came from the UK's share of a fine imposed on a cruise ship for oil pollution. Then again, with a twenty-year timescale, major structural challenges and a £12,000 cost of getting a licence from the Marine Management Organisation just for doing what the government apparently wanted people to do, it will need all the money it can lay its hands on.

The sea is alive with dancing buoys, dinghies under sail, working boats and an annoying jet ski that lets rip ostentatiously as it reaches the end of the harbour speed limit. 'But there is also so

---

[8] 466 species, according to research by the University of Portsmouth.

much more that we are doing besides oysters: eight hectares of restored saltmarsh, seven of seagrass, ten new seabird nesting sites, so much more to our ambition.'

The ambition he is talking about is Blue Marine's Solent Seascape Project, and the 'we' the permanent staff of four who are working on it. The idea is to create a functioning seascape along the coast, where the 'condition, extent and connectivity of key marine habitats is gradually improved and restored in a coordinated way to the benefit of the whole coastline, and of the many resident species there, such as thresher shark, European eel, seahorses, scallops and, of course, those oysters.'[9] I think back to Philip in his little estuary and Sophie in her bay, and just begin to get the first real sense of how things could be if we stopped damaging our seabed, and instead started to mend it.

'How's it going? That Langstone bit,' I ask, and for a moment he is quiet.

'It's early days,' he replies.

Which, as I discover a couple of days later, just may be a diplomatic euphemism.

'I used to be a regular sea-swimmer, but I haven't been for over two years. God knows, most of our group haven't, in fact. Ever since the Sunday that we all mysteriously became unwell after our dip. It's just not worth getting ill for. I mean, I don't know of another sport that can get cancelled because of sewage on the pitch.'

It is a bright October day, I am sitting on the bank above Southern Water's Budds Farm outflow pipe into Langstone Harbour,[10] with Rob Bailey and Kevin Edwards, two activists who

---

[9] www.bluemarinefoundation.com

[10] The inclusion of the word 'Farm' in the name is unfortunate and illusory, but happens to be what it is called.

have made it a mission of their later lives to campaign for cleaner beaches. We are absent-mindedly watching the last of the summer swallows stocking up on insects in this warmest of early autumns, and a huge flock of black-headed gulls sitting at the end of the concrete outflow, presumably waiting for something nutritious to exit it. Something like sewage, possibly.

'Welcome to the outflow pipe for the biggest sewage treatment plant on the south coast,' says Rob. 'This deals with the waste of 450,000 Portsmouth people and their local businesses.'

Rob is a sea swimmer and Kevin a windsurfer. I recognise instantly in both of them the quiet and high-energy determination of a highly articulate 60-something person with a bit of time on their hands and with a wrong to be righted. Much of British conservation relies on people like them. Without this tribe, once widely mocked, nature would be in a considerably worse state than it is.

'At around the same time as he gave up swimming here,' says Kevin, the chairman of the Friends of Langstone harbour, 'I was appointed the first ever Water Quality Officer for a sailing club in Britain.[11] Can you imagine? That a coastal sailing club should need a water quality officer, of all things.[12] It's a strange concept. But then people capsize and go underwater, so we have a duty of care to make sure that they aren't going to fall ill when they do.[13] This goes way beyond the regular litter picks we do along the shoreline. This is about crap and chemicals.'

Sewage has been sent out to sea, directly or indirectly through rivers, for most of human history, and until very recently, spurred on by the view that the limitless sea would absorb without harm the

---

[11] Langstone Sailing Club.

[12] Around ten sailing clubs now have water quality officers (WQOs), or 'Wackos' as they are known locally.

[13] For a graphic description of what actually happens to humans who swim (involuntarily) in deep sewage, you might read an account of the sinking of the *Princess Alice* on the Thames in 1878. Then again, you might not.

relatively small amounts of sewage that came its way. Then, with a population that had doubled in a hundred years, and with a bewildering cocktail of synthetic chemicals starting to come down the pipes as well, the issue gradually became a more pressing one. As one local academic put it: 'there is a staggering list of prescription drugs passed from humans to wastewater treatment plants and into receiving streams, estuaries or oceans by direct consumption, metabolism, excretion or by toilet flushing of old prescriptions'.[14] This situation and technical advances led to water companies treating the sewage through various different methods, including settling tanks, skimming, ultra-violet screening and the addition of other chemicals, before it is sent on its way. These days, the one permitted exception to treatment is in the aftermath of storm events, when the sheer quantity of run-off rainwater will risk untreated sewage going back up into drains and causing dangerous contamination if it is not released without delay. In such instances, and only in such instances, it is allowed to be discharged into the rivers and the sea through outflow pipes like the one we are looking at, the one with the expectant gulls.

Only, that's not how it is actually working. Here, or just about anywhere else in Britain, actually.

In reality, the six-foot diameter discharge pipe at Budds Farm discharges over a billion litres of stormwater into Langstone Harbour each year, or the equivalent of around 400 Olympic-sized swimming pools.

'Southern Water calls them "spills",' says Rob with a wry smile, 'which is what I thought you did with a cup of tea.'

On 9 July 2021, at Canterbury Crown Court, Southern Water was handed a record £90 million fine after pleading guilty to 6,971 illegal discharges of sewage into the rivers and coastal waters of Kent, Hampshire and Sussex, totalling 61,704 hours in all, or the

---

[14] Professor Alex Ford, University of Portsmouth, quoted in Clean Harbours Partnership newsletter.

equivalent of around seven unbroken years. It turned out that they were not just doing this after storm events, but with huge regularity in normal times as well. Just in case you are tempted to think that this was all a bit of an oversight in a large, busy and well-meaning company, it is worth drawing your attention to a section from Mr Justice Johnson's sentencing comments:

> Each of the 51 offences seen in isolation shows a shocking and wholesale disregard for the environment, for the precious and delicate ecosytems along the North Kent and Solent coastlines, for human health, and for the fisheries and other legitimate businesses that depend on the vitality of the coastal waters.
>
> Each offence does not stand in isolation. It is necessary to sentence the company for the totality of the offences to which it has pleaded guilty. But even that does not reflect the defendant's criminality. That is because the offences are aggravated by its previous persistent pollution of the environment over very many years.

The damage to delicate and protected environments was virtually incalculable and, in the short term at least, irreversible, and the Budds Farm works where we are now sitting was one of the main areas. It is little short of astonishing that, in a society based on law and order, no one went to prison.

You might think that this savage judgement and accompanying fine would be the happy end of a sad story, but you would be wrong. That is not how private equity Britain works.[15] The fine-weather discharges continue,[16] albeit at a lower rate, but at least this time they have spawned a little army of watchers, from

---

[15] Southern Water is currently owned by Greensands Holdings Ltd, which in turn is owned by a group of long-term investors representing infrastructure investment funds, pension funds and private equity. (Southern Water factsheet.)

[16] A BBC investigation in the summer of 2023 identified nearly 400 incidents of 'dry spills' from Britain's 3 biggest water companies from the years before.

concerned mothers to drone pilots, water monitors and retired scientists, kayakers and swimmers, all of whom are doing their bit to keep Southern Water on task. This is where Rob and Kevin come in. Working on the premise that one collective pressure group would be more effective than many disparate ones, Rob brought together no less than 25 of these groups under one umbrella organisation, the Clean Harbours Partnership (CHP).[17] To ensure its activities were based in science and evidence rather than emotion, the CHP then promptly raised £30,000 via a crowd-fund and engaged marine biologist Professor Alex Ford from the University of Portsmouth and Dr Tom Miller of Brunel University to supervise a full and regular programme of sampling and analysis, to add to the 'citizen scientists' who were already taking water samples from around 25 locations in the harbour area.

Much of what was sampled was already painfully visible without any analysis, from sanitary towels, plastic applicators, condoms, syringes and earbuds, while the chemical analysis itself revealed something of the hidden underbelly of what is being discharged into our coastal waters: painkillers, contraception, cocaine, MDMA, pesticides and at least 50 other unwholesome concoctions. In one post-storm sample taken by CHP near the Budds Flow outflow pipe, a reading of 380,000 colony forming units (cfus) of E. coli was recorded, which is around 760 times the safe levels as set out in the European Bathing Water Directive; but while eye-catchingly high, it was by no means unique. To be fair, some of this would get through irrespective of whether it had been treated or not; just as sea horses regularly ingest anti-depressants, for example, research has shown that endocrine disruptors in a number of chemicals are now causing some sea creatures to effectively undergo sex changes. After all, 15 per cent of the British population of 68 million take

---

[17] The Clean Harbours Partnership (CHP) is a Community Interest Company (CIC). The website can be found at www.cleanhoarbourspartnership.co.uk

five or more prescription tablets a day and, for the record, a brand new chemical gets registered somewhere in the world every couple of minutes. Many of these chemicals linger for years, often long after they are banned, and, in lingering, are ingested by the local sea life. We cannot pretend that the problem is an easy one to fix. It would just be nice to know that the companies involved had felt that environmental and human health came ahead of shareholder dividends and management reward.[18]

You might also be tempted to believe that this example of red-in-tooth-and-claw capitalism had at least left these companies financially strong, and therefore able to invest into putting right what they have let go unfixed for so long. You might like to think it but, again, you would be wrong. In its Monitoring Financial Resilience Report of October 2023, Southern Water was one of four water companies warned by Ofwat that there was a high chance that intervention would be needed to stop the company falling into financial crisis. In other words, by short-changing their stakeholders, they had ultimately also short-changed themselves. Since privatisation, when all existing debt was written off, the industry has borrowed £53 billion and, at the same time, paid £72 billion in dividends, leaving the cupboard bare.[19] The chances are that it will be you and me, and not the water companies, who will be paying to fix the problem, either through bigger bills or through renationalisation. For fairness, you might also like to think that the situation in Scotland, where the government owns Scottish Water, is better. Alas, you would be wrong again: they just don't bother to measure or

---

[18] For the record, former Southern Water CEO Ian Macauley was paid £1.082 million in pay, bonus, pension and benefits in 2021/22, the year *following* the £90 million fine (*Kent Live* report, 14 July 2022).

[19] Hansard debate on Financial Resilience of Water Industry, 28 June 2023.

report the sewage discharges, which they would find is as filthy as down south.[20]

There is an oft-repeated claim that river (and therefore sea water) quality in Britain is better now than at any time since the Industrial Revolution, a claim that, to me anyway, only stands up to the most selective scrutiny. Certainly, things have improved in sanitary terms, 'but not with respect to emerging contaminants, whilst river quality in catchments with intensive agriculture is likely to remain worse now than in the 1960s'.[21] Things are getting better in the towns but worse in the countryside, and pressures from nutrients and pesticides are likely to get worse before they get better. We need to be sure these days that we are getting angry about the right things, so my suggestion is that we moderate the doomsday hyperbole and concentrate, instead, on the massive 35-year wasted opportunity that could have produced a world-class waste system, and the criminal (yes, *criminal*, as the judge at Canterbury said) behaviour of some of the water company leaders.

'The problem,' suggests Kevin, 'arises from these theoretically highly protected areas not in fact being protected at all.' He waves a hand towards the harbour. 'This has all the fancy letters you could spray on a blue-plaque place: RAMSAR, SSSI, AONB, and yet they seem to

---

[20] A report by Environmental Standards Scotland pointed out that four storm drain overflows spilled waste water into Scotland's seas and rivers more than 500 times last year, and insisted that action must be taken to improve the operation of storm overflows in Scotland and that, in one shocking example, 'the Meadowhead treatment works in North Ayrshire spilled wastewater 365 times on 124 different days in 2023'. So, no evidence that public ownership is necessarily any better than private ownership. 'Storm overflows - An assessment of spills, their impact on the water environment and the effectiveness of legislation and policy', Environmental Standards Scotland, September 2024

[21] 'Is water quality in British rivers 'better than at any time since the end of the Industrial Revolution?' Paper for Science of the Total Environment. Whelan et al. 15 October 2022.

count for nothing. Just inland, there's a little chalk stream called the Lavant that is now one of the most polluted rivers in Britain, because the local treatment plants are not fit for purpose.'[22] He adds that, while 88 per cent of European coastal bathing water sites are assessed as of excellent quality,[23] the UK currently languishes at 72.1 per cent.[24]

It is hardly surprising that those Langstone Harbour native oysters of the Solent Seascape project are not thriving. How could any living thing thrive under those circumstances?

Later on, Southern Water grant me an interview, but only on the condition that it is off the record.

The conversation, during which I begin to feel that I am passing through some out of body experience, lasts an hour, and runs through a long list of why it's all basically someone else's fault. They blame, in no particular order, the age of the network, privatisation, overpopulation, consumers, Ofwat,[25] and, intriguingly, Joseph Bazalgette.[26] When I confront them directly with the small matter of that £90 million fine, they blame the old team for 'gaming the system', previous owners and then the activists for not being able to speciate E. coli. Right at the end they suggest some practical ways that they are trying to improve things and, correctly, how altered

---

[22] A fact that I watched for myself, live, a few months later while I watched raw sewage being redirected from a bowser into that delightful little chalk stream.

[23] European Environment Agency briefing sheet. 3 June 2023.

[24] Bathing Water statistics. Defra. Gov.UK. 30 November 2022.

[25] Water Services Regulation Authority. Ofwat set the investment ceiling for each company, so as to control increases in consumer charges.

[26] With some justification. For it was Joseph Bazalgette who, in 1858, when presented with the two options for a sewage system, one that sent both sewage and rainwater through the same system and one that split the two, (meaning sewage was a far smaller problem during flooding), chose the first and wrong one.

consumer behaviour could help them do so. Reasonably, they also ask me to consider areas where the service has improved over the last 40 years, as well as where it has deteriorated, and to mention the fact that no dividends have been paid out in the last seven years.[27]

The hour is not quite a total waste of time, but it seems a shame that someone senior within the business could not have spotted an opportunity to put across their side of the story, and to paint a positive picture of the future. After all, much of the management has changed, and there is remedial work being carried out, not least a vast storm drain in Newhaven.

Back at Budds Farm, it is a fine autumn day with no discharge apparently taking place. A curlew bubbles its haunting call as it flies low across the harbour towards Hayling Island; three oystercatchers forage busily along the shoreline. Even with all the problems and threats of pollution, Portsmouth is staggeringly lucky to have the heart and lungs of this marine paradise close by.

I tell Rob about my conversation with Southern Water, and he is not impressed: 'Blaming it on stuff that happened 40 or a hundred years ago, that's pathetic. And, anyway, they didn't have to buy it in the first place. The whole justification of the exercise was to make private investment available to improve things. They should be talking about harnessing new technology – you know, things like methane powering housing estates – rather than tinkering with the legacy of the last century. And they make the £90 million fine sound like some minor administrative error; it wasn't. It was a criminal conviction. Gaming the system is OK; lying and destroying records definitely isn't.'

He accepts the points about new investment and initiatives but leaves his greatest scorn for the point about E. coli.

---

[27] True.

'It's just nonsense,' he says. 'We test for E. coli. from human intestines, just like they do. If he's right, then all the seagulls round here must be taking anti-depressants, drinking coffee and snorting cocaine. We might as well just blame it all on the curlews!'

Rob is also scathing about what he sees as the reasons behind the lack of progress: 'You begin with a former government that applied an almost farcical lack of obligation on the newly privatised water companies to spend money on infrastructure improvement rather than borrow it to pay huge dividends to shareholders. You then add local MPs who are desperate not to rock the party boat, and an underfunded regulator who routinely lacks the teeth to do anything other than send out compliance warnings.' As one writer put it, 'governments are like diplodocuses; no matter how hard you stamp on their tails, it takes a very long time for the message to reach their brains'.[28] It appears that, as with the farmed salmon in Scotland, the industry regulator is pitifully ineffective. Sewage in rivers and coastal waters is an area of British life in which it is increasingly difficult to identify any real grown-ups in the room. From the coracle fishers of the polluted River Teifi to wild swimming groups in Kent, entire ways of life are coming to an end for entirely avoidable reasons. 'As a species,' said one headteacher to me a few months earlier on a similar subject, 'we are neither sufficiently engaged not sufficiently enraged.' Exactly. That's why the water companies still do it. We allow them to. And it doesn't help that we don't take responsibility for our own actions, either: presumably we think that a simple practice like not flushing lavatories during the height of a storm is beneath us, or no one has ever told us to do this.

As we walk eastwards towards Langstone village, across some pasture that has been surrendered back to the sea, I wonder aloud what the CHP can actually do about this massive problem, apart from simply draw people's attention to it.

---

[28] *Sea Change*, Richard Girling

'We can be a complete pain in the backside that never goes away,' says Kevin. 'Back in September, *BBC Breakfast* devoted the best part of a whole programme to this issue. They interviewed us but, much more importantly, they interviewed the kids at the local sailing club. They made it impossible to ignore. Campaigns like Feargal Sharkey's are putting the situation in rivers and streams in the full public glare. Meanwhile, we will keep publishing the formal findings of our water testing, keep demanding earlier and better warnings when there are to be storm releases, keep reporting breaches to the Environment Agency. You would think that one day the penny will drop. They will get better at what they do. They will build storm tanks rather than just talking about building them. They will work in a tougher regulatory regime.'

When I ask Rob what keeps him going, a question that I have asked numberless people fighting countless battles over the years, he stops walking and stares out across the coast.

'Optimism, funnily enough,' he says. 'Just think what this place could be like if we stopped trashing it and started looking after it. Think of the social as well as economic value that could be created.'

As he is speaking, a young man walks past us in a dark blue Thames Water sweatshirt, and I catch a few words as he and his friend walk by. It strikes me as odd that the only one I really hear is 'purity'.

He is making quite a statement coming here, wearing that, I think to myself. But at least it's not Southern Water.

Benches of Memory

Littlehampton

Our coastline pulses with memories.

Geological, glacial, evolutionary, military, cultural, they are threaded through it like strands of bright colour through a length of tweed.

And nowhere does human memory present more starkly than in the army of seaside park benches that punctuate our coast. For it is coastal settlements where memories of the dead have been rendered physical in the form of the thousands upon thousands of park benches that stare out in their name from the seafront and away into the English Channel. Seaside and memorial benches, it turns out, go together like fish and chips, or seagulls and ice creams. I have been seeing them, and exploiting them to my own use, in every community that I have visited since Durness at the northern tip of Scotland, all those months ago. If anything deserves to be called the leitmotif of my journey around the coast, I guess it is the humble park bench and, just as we are never supposed to be more than a couple of metres from a rat in our domestic lives, so I am rarely out of sight of the next available thing to sit on.[29]

Benches, it seems, are to be a permanent cultural witness statement to the liminal lives of coastal people. In a culture that at best handles death with an excruciating lack of comfort, given that it is the one inevitable that birth confers on us, benches absolve us from the full process of grief; they are what the candles that we light for the departed become when the flames are extinguished.

Scientists tell us increasingly often these days that the oceans are losing their memory 'as a collective response to human-induced

---

[29] A bench, rather than a rat, to be clear.

warming',[30] so it is somewhat quaint that the species causing this biological amnesia is investing so much energy into ensuring they don't lose theirs.

I won't actually reach 'peak bench' until Bridlington and the little fishing village of Seahouses in Northumberland in a few months' time, but I find Hampshire and Sussex have many thousands themselves. Indeed, on a rainy day in Littlehampton I came across the UK's longest park bench at 324 yards with no less than 9,000 coloured slats; like a car park full of spaces, it took me about 15 minutes to decide where to sit. A day earlier, the entire Selsey seafront was punctuated by regular fiberglass benches seemingly built into the very sea defences, each treasured, etched memory safely muffled by the fleece on my back.

Meanwhile, on cliffs, promenades, esplanades, parks, gardens and estuaries, wherever you look, a bench that got there first will be staring enigmatically back at you.

---

[30] 'The world ocean is losing its memory under global warming'. Article in *National Science Foundation*. 8 June 2022.

THREE FEET HIGH, AND RISING 9

Our Sinking Capital: November

> '"I told you so" are the four least satisfying words in the English language'
>
> Bill McKibben, environmentalist

Occasionally, I try to travel on this circular journey as a child might.

By which I mean that, unlike an adult, a child doesn't need to attribute a name or purpose to everything they see, but instead to revel in the 'triumph of imagination'[1] brought about by the whole experience. It's a different form of travel with different outcomes, where the opening of the mind complements the closing of the notebook, where there are no footnotes, only footprints, no objectives, only memories. The Thames Estuary is a good area in which to do this, and St James' Church in the Kent village of Cooling is the perfect place to start. There will be time enough for science later on.

---

[1] A term purloined from Frederic Gros' *A Philosophy of Walking*, which I found to be quite a useful guide to why people like me walk as much as they do, and what they might want out of it.

Charles Dickens found in the churchyard of St James' his inspiration for Abel Magwitch's dramatic encounter with Pip in *Great Expectations*, and, in the right weather conditions, it is more than possible to imagine the clink of the manacles on the convict's legs and the haunting escape alarm sounding out from the prison hulk across the malarial, mist-bound marshes. In its living history lies much of the reason why the British coast has been such an inspiration to so many novelists over the centuries. From Daphne du Maurier to Ian McEwan and hundreds of others, storytellers have used that liminal edge between land and water to carry their tale more memorably than they might have done on some inland farm or town.

Cooling is two days' walk from my next meeting, which is in the heart of London, and I set off knowing that this will mark the first time that I have headed westwards since Land's End. I revel in the opportunity that walking presents for a slow approach into the capital, step by step along the Thames Path, in a way that makes it travelling rather than commuting, and by which the city reveals itself only very gradually, first through the tall Queen Elizabeth Bridge and then through the glassy office blocks in Canary Wharf that puncture the horizon when I am still a good two hours away from them. This method of slow travel and glacial arrival is in sharp contrast to the modern way of reaching cities, which tends to be either superficially, in a car, or suddenly, by plane. Even an arrival by train discounts any hope of slow acclimatisation. Anyway, how things have changed from Charles Dickens' time. And how they haven't.

At Thamesmead, I sit on a riverside bench and close my eyes for a moment, partly because the long walk has made me sleepy, but also because sound is a different way of understanding the change around me. For sure, Dickens would not have heard the continual 'beep, beep, beep' of reversing heavy plant vehicles, nor the police sirens or the aircraft lining up to land at London City Airport; he wouldn't have heard the sheer variety of human accents and languages that I hear, nor the screech of the ring-necked parakeet.[2] And when I open my eyes, there is much that he wouldn't have seen either, like the giant wind turbines, the office blocks of Canary Wharf or the barges carrying in their yellow containers the accumulated rubbish and detritus of the lives of nine million Londoners. His river was still a thriving, grubby echo of colonial life, where tens of thousands of dockers unloaded multitudes of cargoes out of ships from all over the world, in the busiest sea lane on earth; these days, the shores of my river are quieter, and thick with old hulks disintegrating a little more with each tide, and with unused warehouses sinking slowly into the mud. Also, since around 1850, the Thames has no longer played host to a string of grim prison hulks.[3]

My river is manifestly safer than his. Accidents were frequent and often catastrophic, none more so than on 3 September 1878, in what was one of Britain's worst ever public transport catastrophes, when passenger steamer *Princess Alice* was hit amidships by a coal ship coming the opposite way just downstream from Woolwich and duly sunk with the loss of around 700 lives. It was one thing not knowing how to swim, but quite another to be floundering around in the toxic sludge of sewage that, since Joseph Bazalgette's interventions in the previous decade, now bypassed the Thames further upriver, and was pumped in here in ever greater concentrations. Bodies were unrecognisable within hours, not weeks. Even as a former soldier, even with the passage of time, I find the eyewitness reports almost impossible to read.

---

[2] Opinion is divided both as to where the parakeets came from (pick from Jimi Hendrix, Ealing Studios or a pet shop in Sunbury-on-Thames), or whether or not they are a damaging invasive species. What matters is that there are about 50,000 of them, they are bloody noisy and they are spreading out of the capital.

[3] A few months earlier, in June 2023, Prime Minister Rishi Sunak had his proposal for an asylum barge for 1,000 customers just off London City Airport well and truly slapped down by local stakeholders.

'Nothing we give to the water,' as one writer put it about the Thames, 'is ever really gone.'[4] Nonetheless, my river is cleaner than Dickens's ever was,[5] thanks largely to Joseph Bazalgette and his sewers. These days, dolphins and porpoises put in occasional appearances under the capital's bridges, and seals are seen as far upstream as Windsor.[6] The only mention of salmon in a Dickens novel is on Mr Pickwick's dining table, not, as now, when the odd stray, migratory fish might be found all the way up at Maidenhead. Things are far from good maybe, and micro-plastic pollution has recently become the new pervasive scourge of choice, but we need at least to acknowledge where they have got better.

But the river rolls ever on. The rising and falling of its twice daily tides are the living pulse of the city, defining it, just as it did for Dickens, or Shakespeare, or Ethelred the Unready. Through its wide meanders and out to the marshes beyond Dartford, it passes the eroded soils of the 16,000 square kilometres of England that it drains, and up and down it goes, as it always has, the river traffic of trade, leisure and security that helps keep it all moving. Of London's 35 bridges, Dickens would have been familiar with 11, and many of the vistas I will enjoy in the next three days of walking around it, he would have enjoyed as well. The church bells I hear as I wander my serpentine route westwards beyond Woolwich are the ones that will have called him to reluctant prayer more than a century and a half ago. The raucous energy of this cosmopolitan melting pot still truly links his London with mine. The Thames

---

[4] From *The Way to the Sea*, by Caroline Crampton, an affectionate and detailed portrait of the Thames estuary.
[5] Even as recently as 1957, the Natural History Museum was declaring the river biologically dead.
[6] There is an excellent interactive map of recent sightings at https://sites.zsl.org/inthethames/index.php

remains 'the embodiment of London as a city: powerful, historic and outward-looking'.[7]

Before I get to Greenwich, I am tempted for a moment to take the cable car over from the O2 complex to the Royal Victoria Dock on the northern bank, perhaps just for adventure and to add a different form of transport to the many I have already used, but then I think better of it. I'm not here to ride 90 metres in the air for the joy of it. No, I'm supposed to be here to better understand an existential threat to the city that is so inevitable, so slow and insidious, that virtually no one is talking about it. Apart from the man I'm meeting a couple of miles upstream, who talks of little else.

Because it just so happens that the water level in Dickens's Thames was consistently a good twenty centimetres lower than mine, That matters, because it won't be twenty centimetres for long.

Not only was the Thames twenty centimetres lower than it is now,[8] but this rise is accelerating, and it is predicted to rise by as much as a further 100 centimetres by 2100.[9] Thinking about the consequences of this trend is what Professor Ivan Haigh does for a living, not just in London, but around the world.[10]

To an extent, I have been seeing the effects of sea level rise ever since I turned the corner into Cardigan Bay back in the high summer. But it is not until the South East, where the rising sea is

---

[7] *The Way to the Sea*, Caroline Crampton.

[8] *Will London soon be underwater*? Haigh et al. 27/10/2022. Paper for Tyndall Centre for Climate Change Research.

[9] Or even as much as two metres in the highly unlikely event of the collapse of a major ice sheet.

[10] The argument rumbles on about whether or not London is a coastal city, but the Thames is tidal up to Teddington Lock, 90 miles from the sea, and that means it is prone to the same forces as, say, Brighton or Cardiff, who are more obviously on it. For me, it is unquestionably coastal.

accompanied by a land slowly sinking into the clay, that the situation becomes anywhere near a discussion topic. After all, and with everything else going on in our anxious world, it's hard to get too overwrought about something that is happening at the rate of less than half a centimetre per year. The North Sea sends a tidal arrow into, and beyond, the heart of London, reckoned to be the seventh most vulnerable coastal city in the world to sea level rise, where two million people risk eventual displacement.[11]

The protection of Britain's capital city currently relies heavily on the Thames Barrier, which was built as a direct consequence of the catastrophic storm of February 1953, when over 2,000 people died around the North Sea. Something of an outlier in the sad litany of failed or incompetently managed British infrastructure projects, the barrier has been an outstanding success, both in terms of engineering but also forward planning. Since its completion in 1982, its ten 3,500-ton steel gates (each one the weight of 275 double decker buses) have been deployed 221 times against both tidal flooding and a combination of the tide below and the river above.[12] The designers foresaw, correctly, that sea level rise would increase the number of times it got used, which it duly did, increasing from a couple of times a year to five in its 40-year history, with a freak autumn and winter in 2013-14, when it was raised 50 times. Originally designed to last till 2030, engineers are now confident that it will continue to serve London a good number of years after that, but also that it will increasingly need the support of additional defences which, in the worst-case scenario, would include a new barrier downstream. But then sea level rise on its own is not what really concerns the experts. What keeps them awake at night is the predicted levels of peak flooding.

With the caveat that this is probably not a good party game for those already with properties within the newly affected areas,

---

[11] Earth.org
[12] December 2024 figure.

there is a useful coastal risk screening tool available online,[13] which allows you to observe the slow outward progress of land projected to fall below the annual flood level at any date of your choice in the next 125 years. From Chelsea FC's Stamford Bridge ground to half of Kew Gardens, and including countless hospitals, shopping centres and underground stations, the reality is that much of the city will be more at risk of flooding on a regular, rather than exceptional, basis. And it wouldn't be Britain if we weren't consistently adding to the problem by building new developments, such as the Thames Gateway, further into the floodplain. While not in the same league as, say, Peterborough, Scunthorpe or Bognor Regis, areas the size of Fulham, Bermondsey and Battersea are in the epicentre of this flood risk, which is not just possible, but probable. Maybe I – all of us – should be equally exercised by the many other towns and cities in the UK in the same position without the significant flood protection that London already gets from the Thames Barrier – like Portsmouth, Cardiff, Blackpool and Liverpool – but, as I am in London, I decide to spend three days walking my way up and down the river and in and out of the new 2150 predicted floodline.

Ivan joins my walk on the day before COP28 is gathering in the United Arab Emirates. We set off upriver, up the Embankment, from the London Eye towards Battersea Power Station. I have my backpack, Ivan his little blue pull-along suitcase, trotting noisily along on the pavement behind him like an obedient dog.

I tell him that, to me at least, it is both strange and joyful that someone with his degree of relevant knowledge has not chosen to join the other 81,027 registered delegates creating 200,000 extra tons of CO^2 emissions by attending COP28. With just the King's travelling royal household consisting of no fewer than 16 members, and the Prime Minister's press delegation alone numbering 21,

---

[13] www.coastal.climatecentral.org/map

presumably there weren't any hotel beds left. 'I'm probably better off just doing my job back here,' he says.

'The actual amount of sea level rise in London is still very small,' he begins. 'Under three millimetres a year. But it is the current rate of change that should be holding our attention. It is staggering. We used to talk about the effects of climate change coming. Well, they're right here, right now. And it is impossible to talk about London in isolation: this is a worldwide issue.'[14]

At Lambeth Palace, he gets out his phone and shows me a photograph of a 300-year-old painting of the palace, seen from exactly where we are standing.

'This is quite a good example of the problem,' he says. 'In the painting, the palace is about two metres from the edge of the water. Fast forward to now, and it's more like fifteen, and there is nothing less natural than the building of the Thames Embankment in the nineteenth century that created the squeeze. In building the defences, as we have done up and down its length, we have grabbed more and more land back from the river, created a narrower, faster flow and, in so doing, increased the range of the tides as well as the height of the river.' The natural course of the water in a river near the sea, as this is, is to spread out in sleepy meanders across a wide flood plain, and not to force its way in the quickest possible way to the sea. But the British Industrial Revolution put paid to that notion. Even the hardest of winters would no longer freeze the Thames over, as it did on about twenty occasions between the sixteenth and nineteenth centuries, such is its powerful flow.

For some reason, I still get surprised by the level of emotions that are forced up to the surface, even with world-leading academics, by the fact that these issues are always personal for them as well as professional.

---

[14] No less than 139 cities with populations of over a million, including two-thirds of the mega cities, are situated on coastlines.

'One of the scariest things is that my daughter's kids, if she has them, will probably grow up in a world with no coral reefs.' He stops walking for emphasis. As an oceanographer who dives frequently, this may be a much bigger loss for him than it might be for me, but it is still another gaping hole in the list of treasures that we once had but will no longer.

After a mile or so, he suggests that we find a makeshift classroom so that he can show me the presentation that he has prepared on his laptop; preferably a classroom that serves pints, we agree. The Rose on Albert Embankment fits the bill, and we soon find ourselves occupying the only table not already taken up by an unusually cheerful funeral wake.

He starts at the beginning: 'Sea levels seem naturally to vary by around 120 metres or so over a couple of hundred million years and, to an extent, what is happening now is only part of that, a continuation of what has been happening for the last ten millennia. If you narrow it down to the last, say, one hundred years, the sea has risen by around one millimetre a year; that accelerated to around two millimetres in the 1970s, and then three to four from the turn of the century. My guess is that we are now talking about five, which is an acceleration of around 0.1 millimetres a year. Sounds small, doesn't it? But it isn't. We are absolutely on track for a rise of a full metre by 2100, with a remote possibility of two, if the Antarctic icecap gets going. People get this strange idea that climate change is consistent across the world, and it isn't. It's more like five degrees warmer at the poles than it was.'

While he explains the various possible tipping points to me, I idly look out of the window at the Thames and wonder what two metres of extra water looks like. It looks like a lot.

'The biggest contribution, about 40 per cent, to this rise for many years was glacial melt, half of it from Greenland. That's changed, and it is now the thermal expansion of a warming ocean.' As an illustration of the other influences on sea levels, he tells me that the pace of rise actually slowed in the 1940s and 1950s, probably caused by the growing reservoir of water that was being held back in terrestrial storage by 100,000 dams around the world.

The funeral is in full swing, boisterous and fun. We agree that the deceased must have either been hugely popular or funded the entire booze up, or both. Every now and again, one mourner looks quizzically towards Ivan's laptop, as if it might just provide alternate entertainment for her when the wake finally comes to an end. Judging by the general volume, they have been in here for some time. We don't fit in, but neither are we made to feel remotely unwelcome.

'The point is that we are already committed to this extra rise,' he goes on. 'Mitigation isn't enough. If we stopped all emissions tomorrow, it wouldn't make any difference for at least a hundred years. So we need to plan for what we know. But if we *don't* make that change, if we exceed the 1.5 degrees, as we inevitably will, the one metre could become three or four by the year 2300. Right now, we simply need to concentrate on what we know is coming down the line.'

Through generally doing my learning while I walk, I find I have forgotten the delights of learning stuff while sitting down. It's like secondary school, without the acne, exams or rules. With the aid of additional beer, he narrows the discussion down to the London outside our window, the London I have been walking through for the last few days. The London where I was born, come to think of it.

'Fundamentally, the city is extremely well protected, but there are a number of challenges. Sea level rise is the easy bit. We know roughly what it will be and what it will look like. What we can't accurately predict is how much more the tidal range will vary, how storm surges will develop, how much flooding will come down the river from increased flood events, and how much of a problem the run-off from concreted-over front gardens will present.'[15]

---

[15] The area of concreted-over front gardens in London is assessed to be the equivalent of 22 Hyde Parks: www.londongov.uk

We keep coming back to how lucky London is. With £320 billion of residential property, 56,000 businesses and 700 healthcare establishments all being in the predicted flood zone, it has never been difficult to make a case for investment. 'In the great storm of 1953, over 370 people died around here,' he says by way of example. 'In the equivalent storm of 2013, none did.' I think back to Fairbourne in Cardigan Bay – scheduled to be abandoned to the rising sea – all those months ago, and remember just what happens to communities without provable value.

He explains how the protection for the city gradually developed, with 400 kilometres of dykes and sea walls leading eventually to the Thames Barrier being completed four decades ago.

'I'm always fielding calls from journalists wanting me to give them a disaster scenario,' he says, 'But there really isn't one; not at the moment.' When I ask what is being done about rising levels in the future, and the looming obsolescence of the Barrier, he waxes lyrical. 'The Thames Estuary 2100 Plan is a game changer, so much so that I rather wish I had invented it myself. It is an adaptive plan that bases future actions on real changes in the situation, with decisions being made at the point where most information is available. It is now being used as a model all around the world and it's a great example of embracing uncertainty.'[16]

I remind him about the barrier's original stated best-before date of 2030.

'It's probably fine until 2070,' he says, pointing out that its greatest problem might lie in maintenance. 'At the moment, most deployments are in the autumn and winter, and most of the maintenance is carried out in the summer. If we get an increase in summer storms, that could make the process much more challenging.'

The adaptive nature of the 2100 plan allows for a range of actions at a range of times, from as little as simply bolstering up

---

[16] Thames Estuary 2100 Adaptive Pathway Project.

the sea walls and local defences all the way up to building a fully fledged Thames Barrage (not Barrier) somewhere down around Dartford. The barrage would be permanently shut, with a couple of locks, as opposed to the current barrier that is almost permanently open, meaning that London would become a city with a freshwater river. Whatever happens, it is likely to be extremely expensive, and to change London's relationship with its river for ever.

We have been in the Rose for some time. So long, in fact, that the last knockings of the cheerful wake have given way to the arrival of a sour-looking office Christmas party. It is like a scenery change between two ill-fitting acts of a comic opera.

'I'm afraid we've reserved this table, buddy,' says a man who looks as if he hasn't been less afraid of anything in his whole life. 'It's for our MD.' He throws his backpack provocatively onto the table where my empty glass is sitting, and I notice his name tag on the back, with the name of the estate agency he works for.

'Christmas party?' I offer, lamely, wondering whether to offer a 'buddy' back.

'Yup,' he replies. 'And before you ask, we've booked it all. The whole place. Sorry.' And then, half under his breath: 'Off you fuck.'

When we go back outside onto the Embankment, it is dark. The lights beyond Millbank throw long, fuzzy caricatures of themselves across the river towards us, and the city eases into its end of day routine: the sound of sirens leaks towards us from the middle distance, runners pass with lights in their shoes, cyclists with no lights at all. Eventually, we part ways at Battersea's huge former power station, now apparently London's 'most exciting new shopping and leisure destination'.

'So London is not really the problem, I guess?' I offer, in conclusion. In saying this, I know that I am discounting the fellow travellers of sea level rise like changing migration routes, wader nests flooding, salt marshes being inundated and tidal surges coming into freshwater habitats, but I want to keep things simple.

'Not really,' he says. 'It will be protected come what may. Many other parts of Britain will feel the effect before London does.[17] But no, the real problem is in the parts of the world where they don't have the kind of money that builds huge barriers and sophisticated flood defences. Places like Bangladesh and the Mekong Delta in Vietnam. My worry is that they are continuing to build and build on coasts that are not possibly, but certainly going to be underwater on a regular basis in the next hundred years.'

And off he goes northwards into the foggy night, his faithful little suitcase still trotting loudly along behind him all the while.

Early the next frosty morning, I cross the river and start my journey back towards the sea down the north bank.

London is a young place when you have been tramping the coasts of Britain for many months, and to try to see it through younger eyes than mine, I start my walk eastwards with Kabir Kaul. Kabir is one of a new generation of young naturalists from the north of the city, who is making it his mission to use his youthful enthusiasm to bring others of his age to the love of nature that will be crucial in saving what is left of it. Teased and bullied at school for having the temerity to be more interested in frogs than *Fortnite*, he sucked up the pain and stuck with his enthusiasms. Now, on the cusp of university, he is a relaxed guide who notices things that I simply wouldn't, and tells me things that no guidebook would. Besides, no matter how well you think you know the nature around you, you will always know more if you walk with a local. Here and for a short while in the early morning stillness, the river has almost become a place of silence, with monochrome scullers moving quietly by and sentinel cormorants sitting on each river post. Gradually, life and colour edge

---

[17] To be clear, Ipswich, Hull and Boston already have storm barriers protecting them.

back into the day: an overweight jogger in a Chelsea strip pounds heroically past us on the Thames path, a jackhammer starts its work on the opposite bank, and some Egyptian geese quarrel in the mud below. We engage with anyone who wants to talk, which is almost no one in this busy, breathless city.

Kabir will read geography at university, a good way of bridging the science vs arts divide while he is working out future choices, of keeping as many doors unlocked as possible until the point he wants to start opening them. His knowledge and enthusiasm at the age of eighteen makes a mockery of mine at that age, plus there is steel within him that has already put him on television, and the curiosity that makes him willing to walk in the early morning with a writer like me. There will be jobs in urban sustainability for him when he graduates, of that he can be sure, but whether they are jobs that will pay enough to allow him to live independently here, one of the most expensive cities on earth, is quite another matter. I hope so, because modern Britain is crying out to hear nature stories from people of his age and background.

Since the days when I knew the city reasonably well, it strikes me that melting-pot London has become a place that people go to, rather than come from, which makes it harder to talk to a local. With this in mind, I call one of the only real Londoners I know, my younger son's girlfriend, to ask if she'd like to walk with me down her river that afternoon as my guide. I want to know how she feels about the rising sea, but more than that, I just want to ask her about her city, the city she shares with Kabir and nine million others.

Annie is a proud cockney, so she arranges to meet me near Tower Bridge, at Manze's, for a pie and mash.

'You can't have a sensible conversation about my London without starting with pie-mash,' she says. 'Everything about the place will make sense once you've had one of those.'

As I am spooning up the last of the parsley sauce, I ask her how she feels about her rising river.

'It's the least of my problems,' she says cheerfully, admitting that she's never really given it a moment's thought. Like every young person, she has more immediate things on her mind than something that might make itself inconvenient in a hundred years' time, like the cost of living and employment. Today, what is on her mind is the passing of a way of life.

'Sometimes I feel like the last of a breed,' she says, as we work our way around the twists and turns of the Thames Path. 'The accent is dying out, and the places we come from are being gentrified at a frightening pace. If you want to hear accents like mine, you'd be better off going to Essex or North Kent. That's where everyone went when these places suddenly became fashionable and expensive – Bermondsey, Canary Wharf, Limehouse. That's why they call it "estuary English". From the mouth of a 23-year-old, this is quite a powerful and sad witness statement.

The truth is that everything had changed long before Annie was born. It changed with the creation of Margaret Thatcher's all-powerful London Docklands Development Corporation in 1981, who oversaw the final replacement of a community that once provided a workforce of 80,000 to the inner docks with what was, in effect, a giant business park. In an exercise that was simultaneously visionary and brutal, inspired and divisive, a large part of the financial centre of Britain was yanked, unceremoniously, three miles eastwards.

It's not all bad. People buying million-pound flats tend to demand a nice environment, which benefits everyone. New opportunities have arrived. Crime is much lower.

She agrees when I suggest that the changes have brought good things as well as bad, but laments the ruthless edging out of an entire culture. 'Greasy spoons and caffs,' she says. 'It might not be the healthy lifestyle stuff that everyone goes on about these days, but they were an important part of who we were, and where we went. Now it's just Planet Organic and £4 skinny lattes. I mean, what's that about?' She adds that she can walk down entire streets in her neighbourhood these days and not recognise anyone. 'You've got to dig pretty deep to find the community these days. But it was community that defined us.'

As we walk along the dark of the Thames Path, we hear disconnected snatches of conversation from the hurrying people we pass. 'I said she was being a fool to herself', 'That's plain bollocks, that is', and, 'Me and Steven sent it straight back to the kitchen'.

All of a sudden, she darts down a tiny passage in Shad Thames, leading me down the dark steps to the river and a tiny strand that has been revealed by the receding tide.

'My favourite beach,' she exclaims, eyes sparkling, almost as if she has just arrived in some tropical resort. The tide is the bringer and remover of things in Central London, just as it is in Cornwall or anywhere else, for that matter. Although still only in her early twenties, Annie remembers well the waning culture of mudlarking, the practice of combing the London beaches for lost or abandoned bits and pieces that might have been brought in by the tide or uncovered by the wash of a passing boat. Sometimes, she and her mother bring their own dog down here to the beach. 'She loves it!' she adds with enthusiasm.

In an echo of my time with Ivan a couple of days before, we duck in and out of a few of her favourite pubs, and have a drink in each for politeness and the hell of it. The Old Justice ('they tried to turn it into flats but were forced to reinstate the whole thing back to its original state; I love this place!'), the Mayflower Pub (you're not just on the river here; you're almost in it'), and the Blacksmiths Arms, where we gather with some others for supper.

She may come and go from this city as the years go by, but I sense that she will always find her way back. It is a belonging, a connection within her every bit as powerful as any I have seen so far in my long journey. In every fibre of her body, she remains a Londoner.

And deep down, her river still exerts a powerful draw on her during its final miles to the sea.

Walking on Water

The Broomway, Essex

For most of my coastal journey, land has been land and sea has been sea, and there has been no mistaking the two from each other. But there are still a few places where the two bleed into each other like a watercolour in the rain, and where one might travel from one to the other and back again without ever fully knowing which was which, or where the borderline was.

Forty miles east of London, on the northern shore of where the Thames Estuary and the North Sea become one and the same, is the Broomway, an ancient path that runs for seven miles across Maplin Sands to the island of Foulness. Startlingly, given that it is home to only just over a hundred people, Foulness happens to be England's third biggest island,[18] and for two thousand years or more, until the bridge connecting it to the mainland was built just after the Great War, the only way to reach it without a boat was the Broomway. The tragedy of the island's relentless decline, which began with subsistence farming and ended in its modern role as an ammunition testing facility for the Ministry of Defence, is magnified by the hundred or so people who are reported to have met their end getting onto or off it by its unusual pathway. These days, I suspect that the Broomway's title of 'Britain's deadliest path' exists as much to put some drama in the local marketing opportunity as it does for the promotion of proper caution, although it is undeniable that the churchyard in Foulness contains at least 66

---

[18] After the Isle of Wight and the Isle of Sheppey, which have populations of 140,000 and 40,000 respectively, mainly because both have ample fresh water available, which Foulness doesn't. These days access across the bridge onto the island for non-locals is controlled by MOD contractor QuinetiQ and is granted on just one day a month, and then only if you state that you want to visit the heritage centre.

people who never made it across, and that the journey today still carries real dangers for the foolhardy and unlucky.

In theory, it is an easy enough crossing. The wide route through the flat sands is marked at both ends by the remnants of a made causeway, and then proceeds in a more or less straight line about 300 to 400 yards off the coast, skirting the softer sands inshore. Years ago, the way was signposted with hundreds of what looked like upturned brooms,[19] now long gone, and the new reality is that there are just a couple of tall poles and the crumbling wreck of the *Pisces*, which sunk in embarrassing circumstances in 1982. Sea, sand and sky often merge into each other in the flat coastal light, and visibility can also rapidly disappear through fog or mist and the subsequent disorientation that most often leads to bad decision making. And if you happen to search at night for a light to guide you safely back to the shore, the lights that attract you will almost certainly be eight miles across the Thames Estuary at Sheerness, not on the dark and silent nearby coastline, at which point you will discover at leisure how quickly, and completely, the tide can cover your path. The end comes, apparently, not through being sucked wholly into shallow quicksand, but through the deepening water that flows over your head when already stuck in it up to your waist.

The solution, for the first adventure at least, is to follow a guide.

It's an adventure that promises different things to different people, which becomes clear when our little group coalesce early one Sunday morning around our guide, Kev (aka Thames Estuary Man).[20]

At one extreme is Graham, a fiercely competitive commodity trader from London, who seems more equipped for an assault on

---

[19] Hence the name 'Broomway'.
[20] www.thamesestuaryman.co.uk

the south face of Annapurna than for a walk on the sand, complete with a 70 litre backpack, dangling karabiners and a Garmin SOS emergency beacon; at the other, four youngish goths armed only with trainers, tee-shirts, tattoos and their entire group rations borne in just one flimsy green ASDA bag; in addition, there is Cynthia from Vancouver, who is all in pretty pink for the day before she trips into the wet sand and becomes grey, and two local teenage boys in full combat kit, hoping to find peril in the form of unexploded bombs. In between are the rest of us, including the ugliest dog I have ever set eyes on, a shih tzu called Minnow, with a lilac bow and wearing what looks at first sight like a full immersion suit.

'I give it ten minutes at best,' Nathan, one of the young ammunition hunters, says in a stage whisper to his mate. 'Then it's either going home or it's dead.'

'Dead,' says his mate philosophically. 'It's bloody shaking already, and we haven't even started.'

I get the sense that the presence of the dog coupled with its uncertain future has given their walk extra purpose. Mine, too, possibly.

When we set off, there is a stiff easterly breeze riffling the shallow water under the grey sky, and the light is flat in a way that strikes me as entirely in keeping with the experience. Bright sunshine might somehow place a qualification on any desired sense of adventure, and rain would just be boring and uncomfortable. We are simultaneously out at sea and yet on the land, at risk and yet almost entirely safe. Distance is impossible to gauge in the huge flatness, so I stop trying to gauge it at all, and just find enjoyment in the splashing rhythm of my wellies following the wellies in front of me, who are following the ones in front of them. At this particular set of the tide on this Sunday morning, it is a rather lifeless, birdless space, but far beyond the maypole landmark, hazy shapes that could be ships or buildings loom through the mist. All the life we need comes from the stories that Kev tells us from time to time, and from the conversations that spring up with people we happen to find walking alongside us. The sand we walk on is never more than a few inches high, just as the salt water we walk in is never more than a couple of inches deep. I find it all an intensely satisfying experience. I wanted no more than to taste a bit of life on the blurred margins between sand and water, and between my solid country and the amniotic fluid that surrounds it. Getting to wherever the path leads is not the point for me.

When we get to that point after a couple of hours, it occurs to a couple of us how much we usually take for granted the act of sitting down to rest and enjoy the view, given that we are ankle-deep in water and can no longer practically sit down at all. Instead, we rest by standing still from time to time, as Kev paints in words what has gone before.

As always, the coast is alive with history, freighted with personal stories somehow made sharper by their origins on the edge of our island. When I walked from one end of the country to the other two years before, far from the sea, the stories I came across emerged mainly out of churchyards and local museums; here, they rise up out of every stone. Stories of hardship and subsistence, of wrecked submarines and wicker fish traps, of the great 1953 flood. As I listen to them, I find myself marvelling at how we have so quickly become used to the coast as a benign place of safety and pleasure, where for most of our island story it has been dangerous and relentlessly hard. For many hundreds of years, what I am doing for pleasure this morning was done almost as a matter of routine survival.

As we approach the end of our walk, I try to explain to Graham how I am struck by the otherworldliness of this half land after a year of coastal walking, but it turns out that he's not paid his £25 to chat nonsense with a sensitive writer. He's simply here to tick the Broomway off his bucket list and move quickly on to the next thing.

'The Thames Path,' he says, when I ask him what that next thing is. When I suggest with a smile that the karabiners should come in handy for that, he looks at me with pity.

'Whatever,' he says, as if the saying of it both handily concludes our chat and confirms my idiocy.

Back at the Great Wakering Stairs, it is an undaunted Minnow who leads us in, to general astonishment. Artificial breeding may have created within her frame an evolutionary horror story, but it turns out that the genes of the wolf remain deeply embedded within her. You can take the dog out of the wolf, to misquote the old saw, but you'll struggle to take the wolf out of the dog.

My feet have been safe and dry in waterproof boots; Minnow's have been waterlogged every inch of the way, and I am lost in admiration.

PART 4
NORTHERLY

A WILD GOOSE CHASE: IN SEARCH FOR ABUNDANCE 10

Norfolk: December

> 'I am glad that I will not be young in a future without wilderness.'
> Aldo Leopold

Perhaps nothing, with the possible exception of the wild Atlantic salmon, demonstrates the porous nature of our coast more clearly than a wintering pink-footed goose. Philosophically, at least.

For most creatures other than birds, the border between land and sea is normally where the journey ends: animals, plants and most invertebrates are stopped in their tracks when they get to the salt water, and most sea life can advance no further inland than a turtle goes to lay its eggs. Even humans generally need the artificial help of some kind of craft to go more than a few hundred yards out to sea. Birds alone regard it as merely a change of scenery, and as an equal element.

Of the 633 bird species that have been recorded in the British Isles,[1] around 250 actually call this place home, in that they go on to breed here. Seventy-two species, such as the swallow, migrate

---

[1] British Ornithological Union, December 2023.

here each spring to join the other breeders, and a pleasingly similar 73 replace them each autumn, having done their breeding elsewhere.[2] This latter list is dominated by wildfowl, with geese pouring in from the north as far apart as Canada and Siberia for our warmer conditions, but the pink-footed geese who fetch up here in north Norfolk come almost exclusively from just a few breeding grounds in Iceland and Greenland. Their noisy, boisterous arrival at first light each morning on the wet beet fields around the north Norfolk coast is one of the few remaining ways in which the casual British naturalist can enjoy real, obvious abundance. It is a glorious affirmation that, even now, some things never seem to change. To watch intently 70, 80, 90,000 of them moving around on a three-mile shoreline is, for an instant, to be transported back to a time and a place as yet unaffected by our presence.

We tend to reduce the miracle of bird migration in our minds to a simple and predictable twice-yearly movement of certain birds, the swallow and the cuckoo for example, but it is so much more than that. It is happening, one way or another, every month of the year, at every time of day or night, over every distance and at every altitude between the ground and two miles up. It involves birds that stop to breed here, birds that stop to feed, ones that fly straight over or those that simply arrive in the wrong continent by dint of the wrong wind; even the tiniest bird is equipped with a mix of inbuilt clock, astral chart, map, compass, olfactory direction finder and magnetic reader, not to mention a pair of eyes for landmark recognition. I once spent a year following the travels of a single species, the Manx shearwater,[3] and have never shaken off the sense of awe of these flights, nor of the science that is unlayering for us, bit by bit, their secrets.

---

[2] *Bird Migration* by Ian Newton, a magisterial introduction to what lies behind the extraordinary journeys undertaken in the name of bird migration.

[3] A story told in *Shearwater*, Icon Books, 2021.

Those bird species that migrate do so to be around the most reliable food sources year-round, which, in the case of the pink-footed goose, means getting away from the frozen wastes of its summer breeding fields before the ground is covered by rock-hard ice. Guided by an evolutionary satnav that has inched the memory of the journey southwards generation by generation, century by century, they are hardwired to fly with precision from one tiny area to another equally small one 3,500 kilometres away.[4] These days, and at the dictates of climate change, evolution may well now start to shorten that route on the basis of warmer winters further north, and perhaps one day remove the need for it altogether.

In the journey it has undertaken to be here, the pink-foot has had a relatively easy time of it. In comparison, for example, to the tiny 10 gram willow warbler flying from southern Africa or the bar-tailed godwit, who makes an unbroken flight of 10,400 kilometres from Alaska to New Zealand in as little as eight days and loses half its bodyweight in the process, the pink-foot can cover just 3,500 kilometres in little more than a couple of days. It honks the joyful news of its arrival to a shoreline of locals for whom its presence is as much part of the architecture of the passing year as the buying of the annual Christmas tree or the first bowl of strawberries. For one person, though, it is quite simply the oxygen of his wintertime.

My own nine-month journey has been short on the abundance that I have come to Norfolk to seek out, and the man I hope will unlock it for me is a dark figure now looming in the blackness of the pre-dawn by a layby on the wet edge of the coast road that runs from Holkham to Burnham Overy Staithe. A curious feature of a life largely spent in nature are these hushed, tea-fuelled, dawn and

---

[4] This satnav is generally, but not always reliable, and vagrants appear from time to time well out of the expected range (e.g. in North America). This could either represent some error or the beginnings of some further evolutionary development.

dusk meetings that bring us all together. Much activity in nature is crepuscular, and so too must be its followers.

'You got wellies?' he asks. 'Because you're going to need wellies.' He is right. It feels like it hasn't stopped raining for months. Even the puddles have raised islands of mud with their own subsidiary puddles in them. This is the wet part of a wet season in an already wet landscape, and out there in the half-darkness is a water-world.

While most people continually broaden their horizons as they age, Nick Acheson has done the precise opposite. A life of adventure, including ten years in a Bolivian rainforest, travelling and guiding came to a swift end when two factors in his life collided – the feeling that he could no longer justify, to himself at least, the fossil-fuel habit involved in a life of international wildlife guiding, and the realisation that he could no longer live without the geese of his Norfolk childhood, a transition beautifully told in his book *The Meaning of Geese*. For a time, we just stand quietly on the road and wait, as if for a signal. We are waiting for the geese to move from their overnight roosts behind the tideline to the harvested sugar beet fields behind us.

'The numbers have varied a great deal in the last ten years,' he explains. 'Mainly down round here, but up this year. The method of harvesting beet changed from pulling out the whole thing to chopping it off at ground level, leaving a lump of beet and some foliage for the geese. Easily available food for them. But they risk being considered pests and getting themselves shot.' Climate change is increasingly affecting both the timing and distance of bird migration, together with mismatches in food supply and, increasingly, whether a bird bothers to migrate at all; to an extent, this has always been the case, but never at the speed of change that we are seeing now. But still the geese come. These days, over 20 per cent of the half-million pink-feet that winter in Britain do so right here, a phenomenon that may be helped by the fact that the Holkham estate leave the remnant beet in the ground for a few days before re-sewing the fields. He is about to explain something else when a conversational honking stops him mid-sentence.

'There!' he points over the hedge. Familiarity has in no way diluted his excitement. 'Only 14 or so, but it's the start.' A thin 'V' comes over us in the half light, bulky gander at its head, and they head southwards. Then another 'V', maybe 20, and another. Then a larger formation of 50 or so, slightly higher. Nick explains the family groups and the working of the formations. As he speaks, the volume of birds in the sky increases, like a concerto that starts with a single bassoon and then gradually incorporates instrument by instrument the entire orchestra. We watch silently, as the 'V's become double 'V's and then the double 'V's become like ropes across the eastern sky, rising and falling and pulsing with distant intent and life, silhouetted against the thin sliver of yellow light above the watery horizon. Within a quarter of an hour, the whole sky is punctuated by groups of geese, dotted across the greyness, like hieroglyphs over some vast piece of parchment. It is abundance on the wing, and, for me at least, the accompanying noise is a summoning back of some elemental childhood idea of wildness. It is the flat, side-lit world of T.H. White, Paul Gallico and Peter Scott.

'Maybe 10,000,' he says, when I ask him roughly how many geese are in the biggest group. 'But I don't know. I can only guess, and it is only one group of many. As it happens, the wardens are out counting them this morning, so we'll have a figure later on.'[5]

Other geese are on the move, too, but in much smaller numbers. Knots of compact Siberian Brents fly by us low down in the sky, running a quiet and dignified commentary as they go; a distant din indicates where a skein of greylag is heading inland, too; two brash Egyptian geese fly inelegantly past. Identification in light like this is by familiarity of shape, behaviour and noise, rather than

---

[5] 68,000 geese left Holkham that morning, nearly 200 tons of avian biomass on the move.

detailed observation. Nick exuberantly points out one huge flock of golden plover (a 'congregation') to our south, and another of lapwing (a 'deceit') all around us.

As the sky lightens and the goose flights subside, we walk northwards on the old sea wall and watch the subtle changes of the cast list of this drama: a marsh harrier slowly patrols the outer edges of a field that the team at Holkham have recently flooded,[6] just slow enough to be able to spot an opportunity in the wetness below, and just fast enough not to stall; teal descend into the pond at crash speed, and wigeon are everywhere. From time to time, the raucous descending call of a mallard, more village pond than wild place, dominates the soundscape. I am blessed indeed to be sharing the moment with a skilled local naturalist; things that I would never spot on my own, such as sand sedge and string of sausages lichen, or never identify unaided, like the alarm call of a water rail, are gifted to me in the growing light. It leaves me free not to have to think, but instead to breathe it all in. The act of not thinking is an underrated one in our relationship with nature.

For now, there is abundance everywhere I look in this magical place. It is what I came for.

I say 'for now' because a plan is afoot to run an eleven-mile tidal barrage across the Wash from Skegness to Hunstanton, and which would in time change everything. This is the human way, the Anthropocene shaping of the very geology under our feet as if it is all a macho challenge to our ingenuity, rather than a sacred responsibility for us to protect. Nick is as animated when he talks about it as he is about his geese, but differently. For all the fine words to the contrary, we both agree, human exceptionalism still trumps everything in the natural world, like those beaches back at Hoylake, only far, far bigger. Until the day we are forced to concede

---

[6] The adventure of which hugely successful process you can read in Jake Fiennes' book *Land Healer*.

that they don't, jobs, power, transport, shipping and money will always count for more than mudflats, wildfowl and a few old traditional cocklers.

I think back to Chris Binnie and his planned barrage in the Bristol Channel, and note the apparent inconsistency between my support for that and my instinctive opposition to one here in the Wash. Who am I to judge? In my long coastal journey, I have seen that it is only in the honest resolution of the arguments around proposed projects like this one that meaningful progress can be made, that we can begin to calibrate where immediate human needs should dovetail into the long-term ones of nature. There can be no 'one size fits all' answer. Does the greater good of less fossil fuel use justify the regular bird casualties that will inevitably come from offshore turbines? Does the reliable flow of clean power from a tidal barrage justify the disruption to fish migration and seashore ecosystems that must inevitably accompany it? If I have learned anything on my journey, it is that you have to look far beyond the headlines to come close to understanding the answer. For now, I am convinced that the Wash Barrier has far more to do with the proposed offshore container port, fuelling yet more consumer goods and more haulage miles, than it does with any clean energy it might generate.

We pack up and head back along the sea wall to the road a mile or so away. Nick takes a final look through his binoculars at a mixed group of pink-feet and greylags.

'God! I love you lot,' he says. The way he articulates the words, there is no question that he means every syllable.

For the next few days, I have no plan. I just follow the sound of geese.

I find that I'm happy to walk inland from the coast as well as along it, and to backtrack if necessary so as to follow a lead, and I am almost never out of earshot of the restless, rising, falling, feeding birds; a dozen in a small strip of cover here, a few hundred in a harvested beet field there. Alert to the dangers around them, they will suddenly erupt into the air at some disturbance like students after the last lecture of the week, and then sink back down to resume their feeding once the danger has gone or been discounted.

The sense of abundance that I am starting to feel is in stark contrast to what I found at Sandwood Bay at the start of my journey all those months ago, which is no more than the way of nature and the seasons within which the natural drama plays out. For now, I am in a water-world, a land of low horizons, of skies that bleed into the fields and of treelines that fade into the greyness of the next section of coast. The flocks of wildfowl commuting busily from shore to field and back again prompt in me a sense of almost childish excitement. I want to hear from someone's mouth just once how things were before the downward shifting baselines took over after 1970. 'Ray,' people say to me. 'Go and talk to Ray Kimber at Titchwell. He's seen it all come and go. He'll tell you what it was like.' So I track him down to the RSPB reserve where he is still a volunteer in his eighties, and ask if he's happy to share a few memories. He is. To be honest, he's quite amused that I want to know.

Incessant rain keeps us in the staff room at first, among the welcome detritus of a recent team Christmas lunch – mince pies, chocolate and cake. He tells me that he first came to Brancaster as the junior professional at the golf club back in 1966, a full seven years before the RSPB acquired the reserve. He talks about the changing delineation of the coastline, and of a pine copse that the previous owner unaccountably cut down before handing the land over.

'We used to come down and see flocks of two to three hundred turtle doves back then,' he says. 'This year I've seen just one. Same story with twite; same with snow buntings; same with purple sandpiper.' It turns out that Ray has a soft spot for purple sandpipers, and he returns to them more than once in our conversation. 'There's a bit of climate change, I suppose. A bit of human disturbance. But mainly it's habitat loss. That's the killer, the way we farm. The ones I mind are the little ones, like willow tits and nightingales. You know, the last clutch of nightingale eggs that was ever laid here was run over a few years ago by a Water Authority drag line.'

To my initial surprise, he tires quickly of talking about loss and wants to speak equally of gains.

'Things come and go,' he says. 'We get astonished when a spoonbill comes in to breed, but 500 years ago there would have been spoonbills all over this coast. Then the weather cooled and they went south. Now it's warming and they're coming back.' He talks with pride about his part in the restoration of habitat at the reserve, and how the first little egret showed its face in 1991, the first great white egret twenty years later,[7] and now the cattle egret. When you've watched nature somewhere for the larger part of a century, I suppose you don't get so carried away when the new charismatic visitors fly in, like bee eaters and glossy ibis. At one end of the extinction escalator that is bearing the snow bunting and ptarmigan off towards a dead end in our new climate, lie fresh species flying in from the south. Like many birders, Ray remembers numbers and dates with surprising ease.

'Here's a bit of abundance for you,' he offers, warming to the task. 'In 1973, I can remember my excitement when the first avocet flew in. Ten years later it was breeding. Now, we've got about 700 of them.' He recalls with affection the very first one that over-wintered at Titchwell. 'It was badly injured, and we kept thinking that some predator would get it. But it never happened. It made it through to the next summer and, when it stayed for the following winter, another one kept it company.'

---

[7] Now breeding at Holkham since 2017, and joined by the Cattle Egret in 2020.

I ask him about raptors, knowing that back in 1966 his would have been a landscape of DDT, egg collectors and persecution, and therefore a raptor-free skyscape. 'Amazing,' he says, and ticks off the new residents with the fingers of his right hand: 'buzzards, kestrels, red kite, marsh harrier. Hen harrier if we are lucky. To see one of the Isle of Wight white-tailed eagles come up here and get robbed by a buzzard is something I will never forget.'

As the rain eases a little, we go outside on the slippery boardwalk to see what is around, and I ask him that old optimism question.

'You have to be positive,' he says, just like Rob had said to me down at Langstone Harbour, and Joe in Cardigan Bay. 'And to do that, you have to take pleasure wherever you can find it, however bad things are elsewhere. For me, that might be just gazing over something successful at the reserve that we have all worked on or doing a talk at a primary school to a group of genuinely interested children.' Like many naturalists, Ray quietly despairs of the way many modern parents keep their children tethered to home, and away from the adventure and mischief of long afternoons on the marsh. 'How are they going to learn how to love nature, and help save it, from behind a computer screen?'

We are forced back inside by the returning rain, and Ray heads home to continue working on his project of grassing over his gravel garden. Once he has gone, I put full waterproofs on and head down towards the marsh to see if I can spot a wintering spoonbill that he mentioned. I don't, as deep down I know I won't. But it is no great sadness for me; the light is bleeding away from the sky again, and those geese will be on the move again soon, heading back to their evening roosts.

I sit for a while at the hide, partly to escape the relentless rain, but partly to think through what Ray has said. Much of my coastal journey has been as a witness to depletion rather than abundance, but it turns out that he has planted in my brain the notion that abundance is often in the eye of the beholder, and much of the time hidden from our view altogether. Storytellers, and that is what we nature writers are, have an obligation not only to tell the truth as we see it, but also to seek and describe paths to a better situation. This is particularly important in a world where many conservation charities are unremittingly negative, the better to fill the coffers from a guilty audience. It has taken a man well into his eighties in the seventh month of my journey to explain to me that populations will always rise and fall and that some ecosystems thrive while others struggle. We can quote percentage declines for all we are worth, but how is doing so going to encourage the children around us to learn about, love and protect what they still have?

The winter light washes out of the sky, and I get up to go. The geese are still the reason why I am here, and the secret is to be outside and alert during the shoulders of the day while they are on the move.

Decline is not inevitable. Nature is resilient, much more so than we often think.

One of the ways of releasing this resilience is rewilding. Rewilding is a popular term which is often misunderstood and misrepresented, a trendy fig leaf with which some people cover themselves to sound good while saving themselves the trouble of learning what it all means. However, done properly, it can be transformational. It has a number of definitions, of which the one that Rewilding Britain itself uses is possibly the most helpful: 'rewilding seeks to reinstate natural processes and, where appropriate, missing species, allowing them to shape the landscape and the habitats within'. Of the five principles,[8] probably the one that surprises

---

[8] The five principles are: 1. Support people and nature together; 2. Let nature lead; 3. Create resilient local economies; 4. Work at nature's scale; 5. Secure benefits for the long-term: www.rewildingbritain.org.uk

people most is the one that supports people and nature together, meaning that at its heart is the recognition that it has to work for humans as well. This is really important. If it doesn't work for the local humans, it won't be sustainable in our human world.

Inevitably, it is still raining when I get to meet Dom Buscall in his farmyard at Wild Ken Hill, a place you will be familiar with if you have watched the recent seasons of BBC's *Springwatch* and *Autumnwatch*. In 2019, Dom took the decision to turn 1,040 acres (about a third of the farm) over to rewilding; at its simplest, it involved doing two things, one of which was absolutely nothing. The 'doing nothing' was simply a question of letting all the cereal fields within the new area go back to nature; the extra thing was to add in the 'engineers of change' (a handful of beavers, 45 redpoll cattle, 20 Exmoor ponies and about a dozen Tamworth pigs) and then just let them get on with it.

'The key was getting the baseline survey done,' he says. 'That told us exactly what we were starting with.' We are standing on the edge of one of the old cereal fields, now with the softer edges of an unmanicured hedge and the scrubby grass that begins with nettles, thistles and brambles, and ends with who knows what. 'In this field, we had 16 plant types per survey plot; three years later it has more than doubled to 33. We added nothing. It was all in the seed bank just waiting for the right conditions, and the actions of a few biodiversity engineers.'

Dom explains that the next stage to the baseline survey is deciding exactly what they aspire to see on the place. 'Nightingales, for example,' he offers, 'although it may be that we are just too far from the nearest breeding colony for it to work. Nightjars. Spoonbills.' As he is talking, the berried hedgerows of sloe, rosehip and haw alongside us explode with the sound of a group of fieldfares departing. More abundance, I think to myself.

There are many elements to the work at Wild Ken Hill, from participation in the local farm cluster so that improvements can be made on a landscape scale, to hosting *Springwatch*, to acting as the release site for head-started curlews from RAF airbases.[9] Major projects like the one here are exquisitely reliant on money from the various stewardship schemes that the Department for Environment, Food and Rural Affairs (DEFRA) offer to encourage farmers like Dom to farm for nature as well as profit; if they ever stop, in all likelihood so would he. Over the track from where we are standing is the saltmarsh and, beyond it, the sea. 'Back there,' he gesticulates to the fields that we have been walking through, 'it's all about letting nature get on with it. Here in the saltmarsh, it is the opposite: it is outcome-led and all about control, the aim being to host the raising of as many wader chicks as the wetland can hold.' He says that they have had to become hydrologists, while giving every impression of enjoying that new role. They have built bunds and sluices and, judging by the water rising over the top of my wellies, have done it pretty successfully.

I have given up asking people what the biggest obstacle to their conservation work is – the answer is 'the creaking planning system' just about every time – so I ask Dom about something that has been on my mind on and off throughout my travels, the issue of open access. We both agree that, although it has been campaigned over as a simple black and white argument, it is actually anything but. For him, it is a nuanced discussion that on the one hand acknowledges that the way to get the nation nature-literate once again is to encourage and allow them to get into most wild places without let or hindrance, and on the other that nature has precious few places to hide as it is and probably doesn't need to see them reduced further. Dom likes the idea of allowing access all over the farm, on the basis that he is then allowed to close off bits during sensitive operations, or at

---

[9] Headstarting is the practice of collecting eggs from a site where they are vulnerable or impractical (e.g. by the runways of airbases), and incubating, rearing and releasing them elsewhere to try to establish new and viable populations.

sensitive times of year, such as to dog walkers at ground-nesting bird breeding time. This is an issue on which he calls himself a 'hard line moderate'. 'You've got a responsibility as a landowner,' he continues with an enigmatic smile, 'to find ways of giving people access to nature without violating the heritage. In some ways, I find that I started the project as an environmentalist, and then have gone on to become some sort of social activist.' He is proud to have created ten jobs here where there were recently only four, not least because this makes Wild Ken Hill a sustainable local business. 'We all talk of regenerative farming,' he adds, 'as if only farming can be regenerative. In reality, everything has to be regenerative, whether it's a garden, business or a marriage. Regenerative is just a complicated term for "sustainable".'

And always, out there beyond the sea wall, the eternal sea itself, flat, calm and grey now, sighing its way onto the mud and sand. One of the key future stages in Wild Ken Hill's development will be as a small component of one of the government's flagship Landscape Recovery Schemes, which seeks to create a range of semi-natural habitats including wetlands, saltmarsh, species-rich grasslands and wood pasture heath across nearly 30,000 acres of connected land and fourteen miles of coastline.[10] The genius of this kind of project is that it starts to mend the landscape at real scale, and to a real plan, rather than relying on small islands of unconnected, and sometimes undirected, effort. This one will also go on to implant sea grass and native oysters just offshore, connecting the sea with the land beyond, and with the many similar initiatives I have been visiting for the last nine months. Like Eric's native oysters back in Langstone Harbour that form a tiny part of another project, the Solent Seascape, it is both inspiring and frustrating that so much of this work is going on away from human view. Inspiring, because it is happening all the time in hundreds of locations around the coast

---

[10] Full name: The North-West Norfolk Coast Project.

whether we see it or not; frustrating, because its invisibility allows us to go on seeing only decline.

The default condition of nature is abundance. However, when humans have destroyed it through their activities, it needs humans to bring it back again. That is what is happening here.

As the first flecks of light are threading their way through the tapestry of dawn on my final morning in Norfolk, I walk out along the coast path east of Brancaster Staithe to catch the pink-feet one last time.

Coast path it may well be, but the brooding dark marsh beyond means that I am still a good mile from the sea. Curtains of light rain are starting to drift inland from Scolt Head Island when I hear the first skein of the morning, 24 pink-feet flying down the line of the track towards me and Titchwell beyond. As I raise the binoculars, a shot rings out from a hundred yards in front of me, directly below the geese. I watch for the 24 to become 23, for the bulky body of a goose to drop down into the mud, but it doesn't. No goose falls. Then I wait for the discharge of the second barrel, but it never comes. Only in the scattering of the skein does a hint arise of what has just happened. The shot had been fired by a young lad and supervised by a man who had been wildfowling round here for 40 years. We chat companionably in the half-light for a few minutes, until they head off to try their luck elsewhere, and the last I hear of them is the older man's voice saying that the boy might even hit something next time if he remembers to bring his glasses with him.

I can't pretend not to understand this evolutionary human need to hunt. Our species has hunted for over two million years, so it is going to take more than a couple of decades to persuade us not to. I don't shoot now, but for many years I did, and the long-ago dawn wildfowling was the very best of it. Then one day I suddenly stopped wanting to kill things, and that was that.

All the same, I find I am intensely relieved that the young shooter missed. I think I have become soft during my years of walking, as if in my own journey I am trying to echo that of the geese. They pair for life, and I breathe a silent prayer of gratitude that, in this case at least, each pair bond of that skein has survived to fly another day, and into another beet field.

Survival

The only one-word definition of life that I have ever found is 'survival'.

It follows that it shouldn't surprise me as much as it does that much of what I walk on or under on my coastal journey was once here as a surviving life of its own, defying equilibrium, long before me.

I only have to look up at a chalk cliff, or down at the sand below my feet on a seashore, to get some impression of the sheer scale of life that has gone before me and will come after. Billions upon billions of microscopic plankton compressed into a rising seabed became that cliff that we now subconsciously take to be rock; countless tons of the pulverised remains of marine life joined silicon dioxide to create the beach that we relax on; the sedimentary rocks that we scramble over in Cornwall or Dorset will include an uncountable number of ancient life forms. Life is everywhere, moving, creating and surviving, whether we see it or not.

Rarely is this more in evidence than in the endlessly changing variety of shells on each shoreline that I cross, those collected atoms of a recent life that will, like the plankton in the chalk cliffs, eventually break down and become something else that will serve other purposes numberless further times.

Although my journey has been punctuated by a profusion of razor shells and mussels, scallops and cockles, whelks and periwinkles, I never find the cowrie shell that I am vaguely looking out for, just as I never found a Lego shark or dragon. Maybe I was too busy looking up at the birds, or out to sea at the approaching weather. Anyway, it doesn't matter: these days we are asked to take only photographs,[11] and leave only footprints, so I leave things where they were when I found them. As a child, each beach had been a free-for-all, but things have changed.

Pebbles, shells, saints, gulls, cod and ice cream, survivors all. These are my constant travelling companions.

---

[11] To which we should probably add plastics.

THE SHRINKING CARAVAN COAST: EAST YORKSHIRE 11

Late November

> 'Inhabitants of sea towns surrounded by ghosts of a better time'
> Madeleine Bunting, *The Seaside*

To experience a near perfect metaphor for some of the contradictions in modern British coastal life, I take a dawn walk to Spurn Point, the southernmost spot in Yorkshire, 50 miles north of my Norfolk pink-footed geese.

The thin, sandy peninsula of Spurn sticks, crescent-shaped, three miles into the Humber Estuary, like an arthritic, accusing finger pointing southwards at the heart of one of Britain's biggest container port complexes over on the Lincolnshire coast, opposite.[1] Spurn is an inveterate shape-shifter to the whims of the longshore drift from up the coast and of the capricious storms of the North Sea: sometimes island, sometimes peninsula, always moving. Locals

---

[1] For those who can imagine what such an amount looks like, 62.4 million metric tons of imported products enters the country through Grimsby and Immingham each year, most of it headed for landfill sites via a short stay in our homes and gardens.

call it Britain's newest island after half a mile of it was breached and flooded by a huge tidal surge in December 2013, but the reality is that it has been one many times before and will be many times again. Its shapeshifting calls to mind another new island from near London earlier in the journey, Wallasea, created from three million tons of excavated spoil from the construction of Crossrail, and also the growing beach at Hoylake, all evidence that, here on the island coast, the only constant is change. As here, so on all those previous sections that I have travelled, I confirm to myself that our coast is a dynamic thing, where cliffs crumble into the sea on one stretch and sandbanks pulse into it on others. Perhaps nowhere on my journey defines our relationship with the sea more than Spurn: sea-girt, salt-laden and sand-fringed. Indeed, until 2023, it even had the only permanently and professionally manned lifeboat station in the country,[2] together with the only lifeboatman ever to be awarded the full set of the charity's gold, silver and bronze medals, the first and the last of these on two consecutive days.[3]

It is an artist's world in this luminous land, and in the growing light of the dawn. Black seabird silhouettes drift silently by, the gleam ever-strengthening from the east beyond them in thin layers of orange, purple and red. When the pinpoint of sun finally rises above the horizon, I walk across the 2013 breach close to the low tide sea and beachcomb for a few minutes. New habits die hard. Somewhere deep down, I find that I am still thinking of those Lego black octopuses and sharks.

These days, Spurn is an important nature reserve, owned and run since 1960 by the Yorkshire Wildlife Trust. A few miles out to

---

[2] Now based in a new establishment near Grimsby, RNLI's Spurn jetty was condemned as unsafe, and was going to cost too much, and be too complex, to mend.

[3] Brian Bevan MBE. The story of his gold medal, awarded for the rescue of the Panamanian vessel *Revi* on Valentine's Day 1979, is well worth five minutes of your time and an online search.

sea, the 73 turbines of the Humber Gateway Offshore Windfarm transfer the energy of the near-constant wind to 17,000 households along the Holderness coast. Seventy-five miles out beyond them, in the shallow seas that once formed a landbridge with the continent, is the Hornsea 2 Wind Farm, for now the biggest in the world. That alone powers 1.4 million homes. Inshore turbines pepper the fields to the north, too, a reminder of the glory days before 2014, when Prime Minister David Cameron made the inconvenient discovery that saving the planet could cost his party rural votes, and effectively shut their development down for a decade.[4] Ancient and modern, consumer and conservationist, wild and tame, it is a place that has it all. The close juxtaposition of a vast consumer port and a small volunteer organisation re-introducing a few precious native oysters and sea-grass seeds to the shoreline is striking, to say the least.

While Spurn is famous for many things, conservation grazers, seals, scurvy grass and suffocated clover to name but a handful, it is probably best known as the arrival and departure point of a huge variety, and quantity, of our migratory birds. Looking at a map, it becomes obvious why this is the chosen landfall for so many tiny, feathered pilgrims crossing the North Sea. This is not only on the edge of one section of the great Eastern Atlantic flyway, the route by which Arctic breeding birds make their long passage to Europe and Africa in the autumn, and back again in the spring, but also on the Northwest European flyway, which brings waders and wildfowl to winter here after a summer on the tundra. It is where ornithologists flock in droves towards the time of each equinox, and consequently where some extraordinary sightings are made with regularity. Say 'Spurn' to a twitcher, and they will

---

[4] A decision that, five years later, was still costing British households an *extra* £800 million a year, according to ECIU head of analysis, Dr Simon Cran-McGreehin.

be reaching for the car keys before you can say 'Arctic tern'. But to think it was just about competitive birders would be wrong. It is a deeply loved spot, too, a place of nature pilgrimage in a land of nature illiteracy.

'A pallid harrier,' says the Spurn Observatory's warden, Rob Hunton, when I get back to the café and ask him what his spot of the season has been. 'I had been up at the top of the lighthouse doing something else when I saw it, and I reckon I ran the entire length of the peninsula trying fruitlessly to see it again.' He has noticed a steady decline in bird numbers over the years (although he concedes that the 2023 autumn season has been a good one), but he does not believe that avian flu has been as significant a factor down here as it has further north. 'That could possibly mean that we still have the worst of it to come,' he cautions. Right or wrong, I have certainly seen more seabirds on my journey up the east coast than I did coming down the west.

We look out into the low winter sunlight and talk of dark-bellied brent geese from Russia, of thrushes and starlings that swarm out of Holland in a midwinter freeze, and of the key population of little terns that brought him here in the first place. This natural connectivity with Europe is a salutary reminder of the interconnectedness of everything in and beyond our human ecosystems. The idea of all these extraordinary journeys, we agree, thrills both of us now just as much as it did when we were respectively given our first junior bird books; that tiny goldcrest in a garden hedge, the size of a large conker and the weight of a twenty pence piece, might well have flown all the way from Poland or Finland, and crossed a hundred miles of the North Sea in the process. Show me someone who isn't moved by this, even if only slightly, and I will show you someone who has probably lost the power to be awed by anything in nature, maybe anything at all.

He talks about the latest technology at the observatory, called Motus. It's 'an international collaborative network of researchers that use automated radio telemetry to simultaneously track hundreds of individuals of numerous species of birds, bats and insects'.[5] Tagged birds are tracked by more than 1,200 receiving stations around the western hemisphere,[6] of which Spurn has one, creating the most complete picture to date of where these birds fly, what dangers they face and, therefore, what can best be done to help protect them in the coming years. Any of those tagged geese coming in from Siberia, or waxwings from Sweden, or anything that has been tagged, will be leaving an electronic signature as readable as your favourite novel.

Then I ask him what the biggest influence on the comings and goings of the birds of Spurn is. The warming, rising seas? Plastic in the oceans? The depletion of fish?

'All of the above,' he says. 'The biggest influence by far is the shifting balance between land and sea. It's all about changing habitats, some changes good, some not. As we lose sections of the shoreline, so we lose key habitat for breeding waders. We don't know precisely what will happen. But we do know that the whole coastline will continue to be eroded westwards.' Normally, the intertidal areas will keep up with this until we build defences at the top of the shore, thereby leading to coastal squeeze, as the sea level rises but the intertidal area cannot migrate inland.

In the coming days, travelling from here to Scarborough, 60 miles or so to the north, I will receive a continual lesson in just how fast that coastline is on the move.

I only manage three more miles before my next mug of tea.

If someone from Shakespeare's time happened to have walked those same miles from Spurn to Easington, they would have

---

[5] www.motus.org
[6] Correct as of May 2023.

noticed four villages, long gone now, strung along off what is now the west side of the peninsula: Ravenser Odd,[7] Ald Ravenser, Sunthrorp and Orwithfleet, and even more of these lost villages, maybe 30 in all, stretched like an invisible rope of pearls on the east coast. Communities now under the waves, with names like Horton, Dimlington, Old Withernsea, Monkwike, and inevitable accompanying fables of church bells still being heard to this day when the tide is at its lowest. No one should be surprised. It is a process that has been consistently going on at a pace of roughly two metres a year ever since Doggerland flooded over 8,000 years ago, a passage of time that would have seen the whole coast move a full ten miles westwards. Indeed, if you had asked someone in one of the old Viking settlements how far they lived from the sea, they would have answered you in years, not miles. Four hundred was fine, 30 was manifestly not. I will come across more than one modern resident who still does the same.

Mike Welton, who worked for 30 years at the giant gas terminal to the north of the village, is a local historian with the advantage of having been born here, and therefore being able to gauge the changes for himself. He acknowledges that anyone living on this stretch of coast has an innate awareness of the approaching sea.

'We've always had this strange idea that we can somehow halt it,' he says. 'In the war, they moved thousands of tons of chalk into the middle of the Spurn peninsula to protect access to the anti-aircraft battery at the end; and then they put sea defences on the eastern edge to stop the whole thing moving.' I had seen this for myself earlier in the day, on my walk. Their tumbled remains at the tideline spoke volumes for just how well that idea had worked.

---

[7] Ravenser Odd: the original Humber port, long before Hull, Goole, Grimsby and Immingham.

For a time, he shows me the immaculate albums of photographs he has put together over the last few decades, at once a story of a human community in slow decline and a 'then and now' record of physical change. Two aerial images of the coastline, one taken in 1965 and one 50 years later, show in the latter what almost look like cartoon images of vast bites taken out of the chocolate-cake side of the cliff. He's not worried himself: in Viking terms, he is still hundreds of years from a ducking in salt water.

'This place will see me out,' he says with a smile.

He mentions the great flood of January 1953, the one that killed 300 people in East Anglia and Essex, and caused the Thames Barrier to be commissioned. 'Seaweed halfway up the garden hedge, that's all I can remember now. In a garden that's nearly a mile from the sea.' No one died up here that night, but I sense nonetheless that it was a powerful reminder of what could happen.

'These days,' he says, getting a little animated for the first time, 'when Immingham wants to develop another acre of its port facility, they have to release four acres somewhere else to flood in compensation. So they do it over here. It's bizarre. A Lincolnshire business wants to get bigger, so they flood a bit of Yorkshire. You tell me the logic in that.' Later on, one academic did, indeed, explain to me the logic of that. He said that the idea was to protect cities like Hull, to reduce costs instead of building seawalls and create areas for wildlife. Then Mike adds as an afterthought, with a wry smile: 'It was bad enough that they called us "Humberside" for a few decades.'

This, I start to think, is a good illustration of the Anthropocene in action, where our species is starting to shape and mould the very geology below our feet. Just north of Mike's village is a giant North Sea gas terminal, a critical piece of national infrastructure that will get protection come what may, whether from the Ministry of Defence police who patrol the site or the sea defences on the shore that ensure it is never overrun by the sea. But, as any former geography student will remember from their lessons in longshore drift, one man's protection is the next one's loss. The sea will have its dues, one way or another.

Even with the gas facility just behind it, Easington is a delightful village. Down south, I sense, it would be thriving, with pretty tea rooms and ridiculous house prices to match. But this is not down south. It is right at the landward end of a long peninsula, and consequently on the way to nowhere. The nearby railway line to Withernsea closed in 1965. Four pubs became one, three shops none. Normal story. Like many places that I have been through, it has the vague air of a community simply waiting for the next thing to happen to it.

On my way back to the coast path, I meet two sea anglers loading up their car.

'Any luck?' I ask. It is my normal, annoying go-to question for sea anglers. I can't help myself.

'Five hours. Three whiting,'

'Will you eat them?' For an instant, I realise that at least one of them thinks that I am asking in a roundabout way if I can have them for myself. I hastily correct the impression.

'Not me,' he says. 'He will. But then he'd eat his bloody cat if it wasn't looking.'

It's a funny old thing, that. A seafaring people who have not yet learned to enjoy whiting.

If in doubt about something, ask an emeritus professor.

They generally offer a measured lifetime of accumulated research and wisdom, possibly coupled with slightly more time than earlier in their careers, to explain things to people who genuinely want to know about them. Maybe, too, the generosity is spurred by a wish to make the fruits of their research accessible to a wider public.

A day later, and for the price of a plate of fish and chips,[8] I am getting a field tutorial in the subtleties of coastal erosion. This is a subject that Mike Elliott has been concerning himself with for over 40 years, both at the University of Hull and in a research consultancy, and he is clearly the right person to be listening to. After all, soft clay cliffs, narrow beaches and powerful waves have combined to make this one of the fastest eroding coastlines in Europe, and no one outside the immediate area or academia is really talking about it.

'See those cliffs up there,' he says, pointing north to Flamborough Head. 'That's the last bit of reasonably solid rock until Dover. Everything on the coast between here and there is made up of glacial boulder clay, which is why it erodes so quickly in comparison to chalk and, say, granite.'

The softness of the coast here, he continues, goes back to the most recent ice age, or rather the end of it, when the melting ice left in its wake a huge amount of material that it had dragged along underneath, depositing it on what is now the Holderness Plain as boulder clay, or 'till'. This makes simultaneously for some very fertile farmland and some low but extremely unstable cliffs. By a combination of rotational slumping,[9] topples, falls and mudflows, the cliff line thus edges inland, taking everything with it as it does so.

Protection, where it is afforded, comes in a number of different forms. In Bridlington, it is via a 4.7-kilometre long seawall; at Hornsea, it is a combination of sea wall, groynes and rock armour; at Withernsea, they have tried to make the beach wider

---

[8] Point of order. The fish and chips on this coast are by a degree of magnitude the best I have ever come across. They are reason enough on their own to spend time here.

[9] Geo Forward describes rotational slumping as 'a type of landslide movement that is above the centre of gravity ... due to an increased normal force on the slope, as well as reduced cohesion factor of the soil mass materials.'

by using groynes. At Spurn Head, it is all sorts of things, none of them evidently working. Around ten kilometres, or 16 per cent, of the Holderness coastline is currently under some form of protection.[10]

When I ask him in childlike wonder if all this disappearing land means that Britain is getting slowly smaller and smaller, and the coastline consequently shorter, he smiles and says, 'No. Not really. Because the sediment that is being scooped up here, for example, is washed southwards and probably deposited in the Humber Estuary or further south in the Wash, where it protects the land and towns in the area. Win some, lose some.' Those pink-footed geese I had recently watched at Holkham were starting each day on land borrowed from here. Some of the silt might even work its way over to the Netherlands, due to an anticlockwise gyre current moving around the North Sea, but possibly not a great deal. He explains that the erosion will continue, probably accelerating because of the depression of the River Hull, until it meets the chalk again twenty or so miles west, in about 10,000 years' time, give or take a millennium or two.

We are walking along the seafront south of Hornsea, a small town about twenty miles north of Easington. There is a sharp winter wind blowing out to sea, and the esplanade is curiously quiet. Unlike Mike, who is just getting into his stride.

'You can never protect everything. For a start, at £2 to £3 million per 100 metres, meaning £40 million per mile, it would just be prohibitively expensive. Secondly, you will always just be shifting the problem down to the next section of coast.' We have reached a point at the southern end of the village just in time for him to illustrate his tutorial with a timely example. 'Look at the groynes to your left,' he says, 'and you will see effective protection for a

---

[10] East Yorkshire Coastal Erosion. www.urbanrim.org.uk

settlement of eight or nine thousand people, put in over a century ago, bolstered in 1991 by these ones in front of you. Now look to the right and see what that protection has done to the farm immediately to the south.'

It is a striking example. Everything to our south is about 30 or 40 metres further inland from the seafront, as if someone had taken a giant spade to it. Land gone. Chicken shed gone. Farmhouse gone. Mike tells me that the farmer sued the council, unsuccessfully, for causing the loss of both her home and a substantial chunk of her farm. 'She had a point,' he said. 'When they first moved in during the 1960s, the erosion was minimal, and the sea was 150 yards away. Then the protection was built to their north, and everything accelerated. It is where the legal definition of "act of god" gets really quite complicated. It always ends up with a cost benefit analysis of what is being protected,[11] but that's not going to help the human beings involved.' Then he adds as an afterthought: 'Oddly enough, the safest legal position for a council these days is probably to do nothing, other than to prevent people building in areas to be lost in the next 50 to 100 years.'

There are two more things he wants to show me on this privileged field trip, so we head north in his car for a few miles until he pulls over at the entrance of what looks like some sort of isolated and low-lying industrial complex, hidden behind a hedge. Only an official-looking sign hints of the significance of what lies beyond the hedge.

'That is the Atwick Gas Storage facility, a large chunk of the country's emergency supply. I think it holds about 300 million cubic metres, around 1,500 metres below the ground. Each cavern

---

[11] Normally, according to Mike Elliott, the cost benefit analysis demands a ratio of 12 (being the value of what is being protected) to 1 (being the cost of protecting it).

holds the same amount of gas as 400 gasometers, for those who remember such things. The gas is in old salt strata that have been hollowed out by a process called solute mining. The salt gets pumped out to sea, and the gas from the sea gets pumped back into the now empty cavern.' I tell him that, as an impractical layman, I find the magnitude and imagination of the thing strangely exciting. I mean, who was the first person who actually sat down and came to the conclusion that this was remotely feasible? Sometimes, in what are often days of remorseless bad environmental news, the one thing that sustains me is the thought that my own species' ingenuity may yet come up with workable answers.

A few miles further north, we reach the village of Skipsea, one of a number of settlements where the road quite literally runs out at the top of a cliff, at the point where the rest of it has fallen into the sea. It makes for a spectacular photograph, but uncomfortable living. These days, like many of the local communities, the village seems to revolve around the caravan parks around it, specifically the one down at Skipsea Sands. This makes sense. You can move a caravan inland by degrees in reaction to the approaching sea, until you run out of your own land to do it on. You can't do that with a house. Everywhere I go, I can see abandoned caravan footprints, in the concrete hardstanding and the disconnected utilities. First the caravan, then the abandoned footprint, then the salt water.

Mike is not as wary as I am of standing on the edge of things that appear to be about to collapse, and he calls me over. Through a broken chain-link fence, he shows me a small line of bungalows edging ever closer to the sea beyond. At two metres a year of erosion, I reckon they have got five or six years left at the most, or maybe just one storm event. Unsaleable, uninsurable, this is the human reality behind the geographical phenomenon. To make things worse, the owner is responsible for the costs of the demolition and clear-up when they do eventually come to fall into the sea. For a moment, I think of asking the lady sitting with her dog three gardens down for her story, and then think better of it. How would I like it if some idiot asked me how I felt about my house being on death row?

'The real answer has to be not to build anything in places like this in the first place, doesn't it? Forget mitigation, forget fighting nature, just follow the common sense?'

'Yes,' he says. 'But that's not how humans operate.'

When I get there a couple of evenings later, Scarborough is in its winter clothes, as if it has suddenly woken up that day, discovered all the summer visitors have gone and is now wondering absent-mindedly what to do about it for the next few months. The bones of summer are still in evidence in the kiosks and signs, but the skin of activity has been sloughed off.

Curtains of rain lash the dark seafront of South Bay, and I duck in and out of shelter to avoid the worst of it. I wander past a slew of heroically open, but utterly empty penny arcades, some ice-cream parlours and the Harbour Bar. This is the place that invented the seaside holiday so, vile weather or not, I opt for half an expensive hour on the slots and a box full of fish and chips in some shelter by the harbour. When I get there, the landlady of my bed and breakfast tells me that I am just about the last visitor of her season, and that I can consequently have breakfast whenever I like. I choose 7.30, as there is much to see. The next morning, I ask her if it has been a good season.

'Possibly,' she says, enigmatically, in thick East European tones which have become emblematic of the hospitality of my journey. A few minutes later, she comes back with a perfect pot of tea and says, 'Yes, it's been good,' as if she has had to go back to the kitchen and check the fact on her computer.

Much of seaside Britain simply closes down in the winter, more these days than ever before, so I am profoundly grateful that she is still there for me.

The driving rain has gone and been replaced by a cold northerly wind and leaden skies, so I decide to walk the length of the shoreline from the far end of the genteel north to the limits of brassy South Bay. I have one final destination at the southern edge of town. An optimistic surfing instructor huddles in his empty grey minibus on North Bay, where grey seas crash against a grey wall under a grey sky. Obedient lurchers abound, trotting after striding owners. A tattooed girl on a bench tells someone at the other end of the phone that they can 'fuck right off', and then repeats it more slowly, as if the initial saying of it has given her unexpected pleasure. I find a group of birders among the stacked lobster pots at the harbour trying to photograph a black-throated diver, an elegant but faded bird that is doing all it can to prevent them getting a straightforward shot by diving down for lengthy periods and surfacing in unguessable spots behind impenetrable obstacles. 'What's so special about a cormorant? Why all the fuss?' asks a stout man of his beanpole wife as they walk by.

'It's not a cormorant,' she rebukes him. 'It's a duck. One of those big duck things.'

I double back on myself, and my walk ends just south of the town, in a patch of scrub and grass behind another chain-link fence. For a moment, I just stare across and take it all in.

If I had the sort of mind that worked this way, I could have just about heard the faint sound of a tea dance and of hushed voices on the lawn; I could have just about seen tail-coated waiters bringing jugs of cordial to two elderly spinsters on the terrace and a bellhop transferring a heavy trunk from the boot of an old Daimler to the lobby of a hotel.

I don't have that kind of mind, but it would make no difference if I did, because a few years ago, the hotel simply slid into the sea.

Early on the morning of 4 June 1993, a guest of the Holbeck Hall Hotel looked out of his window and noticed that a large section of garden, a 55-metre section to be exact, had disappeared down the hill, and informed the management. The hotel was evacuated, albeit with strong reluctance from some of the guests even when they could see a large chasm just feet from the hotel. A day later, its entire east wing duly followed the garden down the cliff in scenes that were relayed across the world by a rapt media.[12] Technically, what had happened was that a large rotational landslide had brought a million tons of glacial till down the 60-metre high cliff, and deposited both it and its contents as far as 100 metres out to sea. Humanly, what had happened was that a large part of a huge building had gone for ever, and with it livelihoods and a century-long segment of local history.

Water pressure had built up within an unstable, steep and badly drained slope after 140 millimetres of rain in the months leading up to it, and that is what caused the slip. No one was killed, or even hurt, so the story quickly became just something that regional papers and TV could drag out on anniversaries. The hotel owners unsuccessfully sued the Scarborough borough council, alleging that they had not taken any practical measure to prevent the event. Not so, said the court, who felt that it could not have been 'reasonably foreseen', a precondition within English law for a finding of negligence.

It is a strange feeling standing here 30 years later, as if I am prying on some family's bad luck. I remember it all well from the press coverage at the time, and now I am here, staring out at the anonymous emptiness of a reshaped hill and a large, vanished building.

---

[12] Famously, the hotel's chimney stack collapsed on live television when the then Yorkshire TV journalist Richard Whitely was doing a piece to camera in front of it.

Time and life have moved on and the expensive houses on either side have survived for a new, if nervous tomorrow.

Down below, and all along the coast to the south, who knows what is going on? I suppose that of all the problems that I have been looking at over the last year or so, this at least has the unique claim not to be our fault.

We didn't cause it, not directly at least, but we still choose to go on complicating our lives by pretending that we can stop it indefinitely. We live in a cash-strapped world of noisily articulated priorities, and it is hard to imagine that a larger slice of the national expenditure is ever going to be directed to Holderness.

On another day, a wetter, wearier me might have been tempted to confect the sliding hotel into some clumsy metaphor for the state of Britain's coastal communities, but this is not what happened. Instead, as I walk back northwards along the town's sea road, it dawns on me that I am entering the last phase of my odyssey, and that Holbeck Hall is just the latest in a long line of thousands upon thousands of human stories that have created so much of our modern coast. I think back, for example, to the Broomway in Essex, and to the extent that for so long courage and a tolerance for hardship were almost prerequisites for a life by the sea; to those islands of faith, like Bardsey, thin places where hermits went to be a bridge between their god and their fellow man; and to the beach at Crosby where Gormley's army of a hundred figures tries to tell us what it means to be an islander.

There was more, much more to see, but, having begun the journey seeing only our capacity for destruction and damage, I can at least now start to appreciate the heroic endeavour that had brought us here in the first place, and created the human architecture I pass.

Jet Black in Whitby

North Yorkshire

On the cliff path twenty miles north of Scarborough, a pair of ropes, anchored securely to a post in the ground, disappear enticingly into the unseen blue yonder below.

Enticing is the right word. My Hebridean grandmother taught me that there is always something to be found beyond the metaphorical streetlights, and for the shortest of moments, until I remember how old I am and quite how unlikely it therefore is that the outcome will be a good one, my inquisitive nature and my sense of adventure combine to push me towards the edge to check it out, and maybe even to follow it down. Instead, I wait till I get to Whitby to ask around and discover that it is probably a route for the gatherers of jet to get down the cliff and reach the otherwise inaccessible beach below.

North of Scarborough, the geology of the coastline reverts back from boulder clay to hard, igneous rock, and therefore the intense coastal erosion moderates considerably. The pebble book has returned to my pack for this section, as it holds out promises of continual interest and abundance among the shingles and fossils that I will be walking on.[13] Ammonites, the author tells me, 'can be seen to great advantage in the East Cliff at Whitby'; later, he confides that 'the only substantial deposits of jet in Great Britain' are also there, and urges me to visit the hard shale promontory called Saltwick Nab. I am no geologist, but any long walk along the British coast will be rewarded, through the rocks and stones, with a continual running commentary on how it all came to be shaped as it did, which goes on to explain in part how and why humans settled in the places they did. Bygone geologists

---

[13] *The Pebbles on the Beach*, Clarence Ellis.

would tell me to collect two of each interesting pebble I find, one for cutting open, and one to display opposite it in the cabinet in its original state; but then they aren't carrying a pack full of kit and trying to walk twenty miles a day. So, instead, I collect just the one piece of Whitby jet, which takes me an hour of searching to uncover. When I get home and apply the test,[14] I discover that it is, in fact, a lump of coal.

For a while, I get a little obsessed with jet, not least because it happens to provide a delicious metaphor for our own miniscule tenancy of this planet.

Some 180 million years ago, when Pangea was beginning to break up, dinosaurs still roamed the place, and when our ancestors were simple egg-laying monotremes,[15] the land behind Whitby was tropical forest covered in something pretty similar to a modern-day monkey puzzle tree.[16] When the trees fell, they would duly wash down the rivers to the coast, and then get covered in successive layers of sediment that compressed their fossilised remains into thin bands of hardened black stone, now prized as jet. Humans have been wearing jet jewellery for 4,000 years, but it was only in the Victorian age that it was found to be a rather fashionable expression of mourning that the monarch of the day seemed to expect of society women; at one point, the town boasted 150 jet miners and a further 1,500 people working and selling it, including in 200 specialised shops. By 1875, the fashion was dying away, as were the jobs, and it wasn't until over a century later that it became popular again, fuelled in some part with Whitby's renaissance as a destination for goths.

---

[14] The test for jet, according to Ellis, is to rub a corner of it onto a piece of pale, dry sandstone. If it leaves a black mark, it is coal; if brown, it is jet. You are probably glad you asked.

[15] Think duck-billed platypus, and then be imaginative.

[16] *Araucaria Araucana.*

These days, the monkey puzzle lives on in the wild in Chile and Argentina, 7,500 miles away, and is only sold in these parts as a slightly frowned-upon ornamental exotic that just happens to have been right here 200 million years before us, an invasive that predates the merest suggestion of our species' existence.

Of the town's five museums, only the Museum of Jet is open on this late January day, but even its undoubted delights are only delaying the real reason I am here.

For Whitby just happens to be at the southern end of a 40-mile stretch of coastline that has recently provided, if it is possible, even more controversy than the sewage in Langstone Harbour, or the salmon farms of Sutherland, all those months ago.

DEAD CRABS ON A TROUBLED SHORELINE: NORTH YORKSHIRE AND TEESSIDE

12

Late January/Early February

'One of the penalties of an ecological education is that one lives alone in a world of wounds'

Aldo Leopold

It started on 26 September 2021, when Hartlepool fisherman Stan Rennie noticed a couple of dead lobsters in his quayside creel.

For a man who had been fishing for 40 years and felt that he had seen it all, it was an unusual sight, but certainly not a cause for alarm. For the next few days, he noticed more dead crustaceans in the creels, as did his colleagues, and an unusual 'ginger' colour to the water. Then, on 4 October, after a bit of rough weather, local bait collectors reported an unusual number of dead and twitching crabs on the shores of Bran Sands, just where the Tees enters the North Sea, a couple of miles from Stan's creels. In the days that followed, more local shell fishermen also found dead crabs and lobsters in their creels, and a vast number of dead or dying crustaceans, mainly crabs, turned up all along a 70-kilometre stretch of the coast of north-east England, from Whitby in the south all the way around to Seaham, a few miles north of the Tees Estuary. At its worst, the dead crabs were piling up in thick layers at the high tide mark, most noticeable along sandy beaches, such as the four-mile stretch between the mouth of the Tees and Redcar, to the south. And not only crabs; lobster, razor clams and shrimps were affected, too. And not just for a short spell; one of the country's long-term environmental monitoring sites just down the coast at Staithes showed a massive drop in the barnacle population at all levels of the shoreline over a year later. In fact, on the lower- and mid-levels, there were none at all.

The acute stage of the phenomenon, which made the national headlines, lasted about two months and then happened again six or seven months later. As much as anything, the story seemed to give further credence to the idea that we were all living in some new and anxious world of man-made disease, and that our careless past was fast catching up with us. After all, Bran Sands lies downstream from Middlesbrough at the southern end of the Tees Estuary, itself largely agreed to be the most polluted waterway in Europe. On its south bank lay the vast 4,000-acre site to the centre of the planned new freeport; full of the waste of old coke ovens and multiple other chemical hazards and concerns (not least of which were two of biggest deliberate explosions ever seen in Europe on its western edge),[1] planning had been granted with a much quicker timescale than similar redevelopments elsewhere.

The Tees at Middlesbrough is a far cry from the stream that chuckles off the side of Cross Fell, where it rises in the Pennines, that thunders over the cataract of High Force a few miles downstream or into whose clean waters I had once dipped my hot, grateful feet one spring Saturday below a juniper thicket near Middleton-in-Teesdale, at the end of some 25-mile walk. To an extent, the main punctuation

---

[1] On 1 October 2022, the Oxygen Steelmaking Plant in Redcar was demolished in one of the largest single explosions in Britain in 75 years. Fifty-four days later, the blast furnace went the same way, and the last traces of a local business that had once employed 19,500 people evaporated into dust.

marks of my coastal journey have been the river estuaries I have crossed, from the Laxford right at the start of my journey to the Mersey, the Rheidol, the Lim, the Lavant, the Thames, the Nar and the Humber; places where humans have congregated for millennia, made markets, built bridges, started industries and created mess. Alternately sewer and thoroughfare, harbour and meeting place, the widening river mouth in each coastal town I have walked through has been, for me, its beating heart. From early industrial times, the Tees has embraced it all – mining, shipping, rubber, steel and chemicals, to name but five – in the pursuit of a living. As one scientist put it to me, for context: 'the whole Industrial Revolution was based on screwing the environment. Now is no different. It just has more paperwork'. And these days, the silent witness to that old industrial embrace lies heavy in the sediment below the waters flowing their way into the North Sea. Undisturbed, it is relatively harmless, but it is rarely left undisturbed.

Experts initially agreed that one – or a combination – of four different factors could have caused this die-off: a disease pathogen, a harmful algal bloom, chemical toxicity or the side effects of dredging in the dynamic estuary.[2] And it just so happened that the Tees Estuary was the chosen location for one of the Conservative government's much-vaunted freeports, a scheme that would create employment for sure, but one that nonetheless would also involve industrial-scale dredging of the seabed and thus the possible release of toxic chemicals that had lain within it for decades, most notably a solvent used in paints, rubber and dyes by the name of pyridine.[3]

To this day, no one can say with complete certainty what caused this mass die-off. The initial DEFRA report swiftly concluded that

---

[2] Other rejected causes included extreme weather events, ruptured undersea cables and parasites.

[3] Chemical name C_5H_5N, pyridine is known to be in the Tees sediment in large, historical quantities and is a known environmental contaminant.

'a naturally occurring algal bloom' was the most likely culprit, not unreasonably, as there was a satellite-visible bloom off Teesside at the same time. It was a view that, as we shall see, they would change completely in their later, fuller report. Also, by conflating offshore catch figures (which hadn't been much affected) with the onshore ones, the government convinced themselves that the loss of income for the fishermen was less extreme than it actually was. Local fishermen, whose livelihood had been temporarily torpedoed, came to a different conclusion, and initially believed that the die-off was directly linked to toxins, especially pyridine, released into the Tees Estuary by a change in dredging activities, activities that would be compounded by the construction of the new Teesside Freeport, a 4,000-acre giant capable of handling the largest container ships of all, and creating some 18,000 much-needed jobs in the process.[4] Critics of the fishermen inferred that, in order to get compensation, they needed first to find a human culprit, and that the pyridine release was the only practical way they would get it. In other words, they would say that, wouldn't they.

On 8 February 2022, ITV News approached a marine biologist at Newcastle University, Dr Gary Caldwell,[5] for his views. Later on, Caldwell was invited to a meeting of the fishermen in Whitby, and introduced to an employee of the Fishmongers' Company, a livery company that had agreed to provide a small amount of funding for some independent research into the die-off. To cut a long story short, Caldwell was quickly asked to work on the report himself, with one of his PhD students.

---

[4] Teeside Freeport figure.

[5] Senior Lecturer in Applied Marine Biology, Newcastle University. For the sake of full disclosure, Dr Caldwell is a member of the Green Party, a fact he told me without prompting. His low-key activism forms a large part of the attempts to discredit him.

Caldwell concentrated on one simple question, which was whether the concentrations of pyridine identified by the initial report were sufficient to damage crabs. This was to test the DEFRA position that 'the amount of any chemical needed to cause a mortality event of this scale would need be huge and could not have escaped detection in the extensive sampling'.[6] Crabs were gathered from well outside the affected zone, and then subjected, under laboratory conditions, to the relevant concentrations of pyridine.[7] Walking along the sea front at Hartlepool, he explained to me what he found.

'To begin with, the crabs in the higher concentration tanks simply went ballistic, somersaulting and racing around. We were astounded, as it was absolutely not what we had expected. After that, they all settled on their backs and died, with the ones at the upper end twitching much as the ones did that they had found on the beaches. The research not only showed that pyridine is highly toxic to crabs but, with computer simulations of North Sea currents and tides, that the pyridine would have been transported along the affected coastline within hours, in sufficient concentrations to kill off approximately 10 per cent of the population within three days. Furthermore,' he adds, 'whilst we know that pyridine is soluble in water, we also believe that it binds itself to molecules in the sediment, which enables it to travel long distances.'

'But they hadn't actually started the main dredge for the freeport at that stage,' I put to him. 'The report makes that clear. So how come there was any higher concentration of pyridine there than normal?'

'It wasn't the freeport,' he said. 'It didn't have to be. In late September, there had been a landslip into the Tees. Then, when one of the maintenance dredgers was out of commission, they

---

[6] DEFRA Press office blog, 30 September 2022.
[7] *Determining the toxicity and potential for environmental transport of pyridine using the brown crab (cancer pagarus)*. Eastabrook et al. November 2022.

brought in a much bigger dredger to continue the work. The dredger brought in was called UKD *Orca*, and it began work on 25 September, the day before Stan Rennie found the dead lobsters in his creel, close by. Maybe it's coincidence; maybe it's a factor.'

Caldwell duly submitted his report, little knowing the storm coming his way once he had done so.

In the afternoon quiet of the Pier Hotel in Whitby, fisherman James Cole has laid two folded paper napkins lengthways on the table.

'Here is the coast going north. And here', he points at the second napkin, 'is the coast coming down towards us in Whitby.' With the help of a sachet of tomato ketchup and one of HP Sauce, he explains to me the north–south longshore drift of the coast that he believes brought the pyridine down to his own fishing grounds. Then, with a teaspoon, he demonstrates the direction of the flooding and ebbing tides. 'It's so predictable,' he adds. 'None of it ever changes.'

Outside the window of the pub, the greyness of an early February evening is settling on the harbour. On the summit of the hill beyond, I can see the ruins of the giant abbey. The wispy clouds around it give off an air that once also inspired Bram Stoker to create *Dracula* back in 1890 in Mrs Veazey's guesthouse at 6 Royal Crescent.

Cole charts his 40-year journey as a fisherman. 'I came from a line of trawlermen,' he says. 'Grandad started off with crabs and lobster, before progressing to white fish, which my dad continued. With all the quotas and that political bollocks, I went back to shellfish. Bought a smaller boat; cut fuel costs by a factor of ten. Worked with my son and another lad. It's worked well for us, by and large, weather dependent, of course.' He explains how he tried scallops, but gave up when he realised just how much he was damaging the seabed. 'That's what no one seems to understand,' he says. 'Not all fishermen are trashing the seas. Not by a long chalk. Why would we, when we need it to be healthy for us to make a living?'

When I ask him how much his business has been affected by the die-off, he is sanguine. 'I just went further out,' he says. 'Incurred higher costs; spent longer getting less, but we survived, and it's coming back now. We knew that it was a crisis within the first five or six days. The *Orca* was there for about two weeks, during which time it worked round the clock and took more spoil out than the locals did in a year. After that, it was just a question of waiting for the poison to work its way down the coast, which it did. The poison they are now denying.'

'Who is "they"?' I ask.

'The government. Its agencies. Conservatives. Look, I'm a business owner, and that's who I used to vote for. Not any more I don't. I'm a green activist these days. The joke is that people like me were never against the freeport as such. This area badly needs the jobs. Just that it shouldn't be at the cost of the environment.' He takes a mouthful of tea. 'If it was the pathogen they keep saying it was, Teesport has become the new Wuhan!'

'How so?'

'Because it has jumped species, just like Covid did. Lobsters. Barnacles. Birds. Kelp. If it was that dangerous, why didn't they warn the rest of Europe and close down the Tees until they had sorted it?' He smiles with a weary acceptance that there isn't a logical answer.

'But was it about compensation?' I ask. After all, that's what they had all been telling me.

'Was it hell,' he laughs. 'We've never asked for a penny. Anyway, the government ruled it out.[8] Look. I'm 55. I can go and stack shelves at Tesco and make as much as I make now. But that's not the point. The sea is for all of us, and for all generations. What about the next person, the next generation, that wants to come and make a living from the sea? What if it's not there for them? Don't

---

[8] May 2023, DEFRA

people like me have a responsibility to make sure it is? Don't we live close enough to it to be listened to when it goes wrong? We don't just go out there for the money, but also for the camaraderie, the freedom, the outdoors and our mental health.' He pauses for some more tea, looks me in the eye, and adds, 'There's two things that worry me still, whatever the cause eventually turns out to be. The first is that we are all just sitting here waiting for the next time. It's like Covid. Once it's visited, you're just waiting for the doorbell to go again.'

'And the second?'

'The crap they've given Gary is unforgiveable.'

The official expert report in early 2023,[9] commissioned by DEFRA after Dr Caldwell's report, looked exhaustively at all the possible causes and came to no firmer conclusion than that 'a novel pathogen is considered the most likely cause of mortality'. Freely accepting that there was no direct evidence of such a pathogen and accepting that there may have been more than one cause, the committee had arrived at the conclusion they did through the act of finding the other possible causes – an algal bloom, dredging or poisoning by the release of the industrial chemical pyridine – either 'very' or 'exceptionally' unlikely. Read in isolation from the political behaviour that followed, it comes over as diligent, even humble science in its acceptance of there being no smoking gun and no definitive answer. Indeed, the report finishes by stating that 'it is also possible that several of the stressors considered in this report operated together to degrade the marine environment and lead to the unusual mortality'. Meaning that they didn't really have a clue.

---

[9] 'The Independent Expert Assessment of Unusual Crustacean Mortality in the North-East of England in 2021 and 2022'. The report was produced by a committee of thirteen experts.

This report was produced by a committee of experts chaired by DEFRA's Chief Scientific Adviser, Gideon Henderson, and with input from the government's Chief Scientific Adviser, Sir Patrick Vallance. The list of the twelve other members reads like a Who's Who of the British marine biology world, with expertise in crustacean biology, marine eco-toxicology, marine pollutants and chemical dispersion to name but four, and runs to 66 pages of careful scientific analysis, with an executive summary that begins: 'Overall, the panel was unable to identify a clear and convincing single cause for the unusual crustacean die-off'. The furthest that they were prepared to stick their necks out was in saying that it was 'as likely as not that a pathogen, new to UK waters, caused the unusual crustacean mortality'. Instead of ruling other possible causes out, it simply accords them with those degrees of unlikelihood. In relation to the points raised by Dr Caldwell's report, they cite examples from USA and China where twitching was linked to disease (and not poison) and go on to dispute his central findings about the reaction of crabs to specific levels of pyridine. As far as dredging is concerned, the report divides this into *capital* dredging, which 'generally excavates geological or historically accumulated sediments to create new or deeper channels', and *maintenance* dredging, which simply removes 'recent infill material' in the original channel. The panel found it *exceptionally* unlikely that capital dredging on the Tees caused the die-off, and *exceptionally* unlikely, as well, that the *Orca*'s activities had caused it.[10]

In the wake of the report's publication, Conservative Tees Valley Mayor Ben Houchen went deeply political.[11] 'It was the aim of these

---

[10] In a further paper (*Why there is no evidence that pyridine killed the English crabs*; Ford et al, Aug 2024. Paper for Environmental Science), three of the original report team double down on the panel's findings. While unconvinced, I do not disagree with their reference to the toxicity of the political debate around the issue.

[11] To be clear, I invited Ben Houchen, through his official website, to be interviewed for the book. I never heard back.

Labour figures and activists to undermine this vision for their own short-term political gain. They were willing to risk a generationally transformative development in a deprived area simply to undermine a positive Conservative objective.'[12] He attacked Dr Caldwell personally on social media on the basis of his being 'an activist', that his modelling and methodology hadn't been made public and that his paper hadn't been peer reviewed. (I have learned over the years that expert academics generally find out that an activist past, or present, will be dragged up by big business, and their independence questioned, if their research is endangering someone's profits.)

Shortly afterwards, his employers, Newcastle University, were approached and asked to distance themselves from Dr Caldwell's research and to consider sacking him, demands that were rejected out of hand. Then the online abuse started, often on private Twitter accounts that Caldwell only saw when colleagues sent them on to him as screenshots. Then the phone calls started, both to Caldwell's mobile and to his office phone, always showing 'number withheld' and always making threats against either him or members of his family. Who these people were, Caldwell hasn't got a clue,[13] but in a way that's not even the point; in an age and place where arguments are routinely conducted in black and white, and at full volume, it has become so easy to make your own point by reaching not for the facts, but for social media and then articulating something that would have you locked up if you said it aloud to someone's face. Life for the keyboard warrior has never been easier, their unpleasantness and hurt never so easily distributed. Other research funding dried up and it wasn't long before he realised that that chance interview on ITV News had utterly changed his life.

---

[12] Interestingly, no one I met in the research for this chapter admitted, when asked, to being a Labour voter.

[13] For clarity, Dr Caldwell made it clear to me that he did not believe that Ben Houchen or his immediate team was in any way behind the online bullying.

'At least my son now thinks I'm quite cool,' he says to me with an enigmatic smile. 'Which he certainly didn't before.'

As it happens, Gary Caldwell isn't the only one getting anonymous threats on account of activities on the Tees Estuary these days.

Ex-career police officer, lifeboat crew, pint-puller, cleaner and occasional marine mammal medic Sally Bunce has had a number and shakes them off like a dog and its fleas. Outwardly, at least. 'They only hurt if you keep quiet about them,' she laughs. 'These days, they are probably just a sign that someone like me is doing their job right.'

Her 'job', in this respect, is twofold: to campaign for proper research into how the recent pollution on the Tees might be affecting porpoises, dolphins and particularly seals. And to be the Green Party candidate for the upcoming Tees Valley mayoral election. When I speak to her, she is preparing for her first hustings. 'I've got no money,' she says cheerfully, 'no manifesto, no helpers, no hope. I just want to use the election to draw people's attention to what is going on in their valley.' As with others, she is scathing about the lack of attention that Tees people have had from central government and the national press. If you ever wanted a walking example of the courage, knowledge and persistence that it takes to hold recalcitrant and secretive public bodies to account, then Sally would be the candidate from central casting.

She has a particular soft spot for Common Seals,[14] which, with the occasional perversity of species names, aren't really that common at all, at least not like the greys. 'Something big has been happening to them since early spring of 2022,' she says. 'And we

---

[14] Common, or harbour seal. Rather smaller than the more prolific grey seal, with a small rounded head and V-shaped nostrils. Being at the top of their food chain, seals tend to accumulate persistent pollutants and pass them on to their pups, who are less equipped to deal with the concentrations.

should be very interested, not least because they are a designated indicator species, sharing much of their diet with humans. What they eat is what we eat. So when they get a strange mouth-rotting virus, when the pups are weaning at lower than their birth weight and then dying, we should be asking science to find out why this is happening. But we are not. Then some twenty harbour porpoises wash up on the beach, and suddenly there isn't any money to do toxicology reports on them, however much I beg DEFRA to do it. And neither are we asking why dolphins, whose arrival around the turbines off the beach at Redcar you could set your watch by, now simply avoid the area altogether. These days, they are just moving from the Farnes to Whitby and giving us a wide berth.

'It's too late to stop the main dredging for the Freeport,' she explains. 'And therefore too late to stop the 1.8 million cubic metres of sediment that has already been dropped out at sea. But surely there is enough uncertainty about what remains to be disturbed to pause it, and make ourselves certain. The current testing regime simply isn't fit for purpose: a surface scrape every three years or so and you're done. And anyway, the dredging safeguards are only as good as the companies working with them and the authorities enforcing them. And with most government agencies pretty hollowed out by austerity, you'd need to be some kind of optimist to believe that it is all being done to the letter of the law.'

Sally Bunce is a well-informed and well-read activist, and she readily admits she is not a scientist. So when I ask her what she thinks caused the die-off, she prefaces any answer with the health warning of the highly interested amateur. It is only when I push her that she first mentions the 'P' word.

'Pyridine,' she says. 'I've got absolutely no doubt that it is.' And then she adds, 'They get away with calling us conspiracy theorists, as if the insult is somehow going to strengthen their case, or make us go away.'

Sally discloses a pattern here. It has been a recurring trope all the way round my long coastal journey, this feeling that the good guys, the careful, fixed-gear, inshore fishermen and women consistently feel that they are paying the price for past mistakes, while the big factory ships over the horizon, the beam trawlers and scallop dredgers, get away with environmental murder. It was thus in Kinlochbervie and it still is here.

There's a second repeating theme, as well, one that I have also been noticing on and off all the way round the island. It presents as no more than a background hum which, only if you listen hard, then discloses a slow ebbing away of hundreds of small pockets of precious human coastal culture, a decline that no one can articulate as well as a man sitting in an armchair when he should be catching whiting out at sea.

This is not to say that people don't have to change and adapt. Of course they do. We all do. But it would nonetheless be nice if their opinion was sought on the way there, and their worth and their contribution to the community both appreciated and recompensed.

I finish my Hartlepool walk at a small terraced house on the Town Wall, overlooking the old dock.

Stan Rennie is sitting on his armchair, watching daytime television, the same Stan Rennie who saw those first dead lobsters on 26 September 2021. His professional life has rather fallen apart since then.

'This'll be my last season,' he says. 'They tell us the shellfish have come back to Hartlepool, but I'd like to see where they get that from. You see, the night that die-off started, other stuff started as well. Within days there was no life at all. The moss on the sea walls died. The seaweed died. The kelp beds died. Even further out, the marine life, like squat lobsters, they died, too. Razor clams were washing up on the beaches around Whitby in their thousands.

The walls were so clean you would think that they had been power washed. Does a pathogen do that? I don't know. The only thing that seems to have been resistant was the hermit crab.' He shows me a video on his phone of a lobster twitching down in the dock below his house.[15] 'I'm 62, anyway,' he adds, as if that makes it all slightly better.[16]

Over lemon drizzle cake, he concedes that it is at least less bad than it was. 'In 2022, instead of going ten minutes across the bay to do my fishing in the bay where I have worked all my life, I was going two and a half hours up the coast to Seaham. The guys are still going out today and getting next to nowt.' Almost as if to reinforce his point, his phone goes; it's a call from a friend who's been out at sea for the morning. 'Three lobsters, four crabs, five velvets.[17] That's all. A bloke's not going to feed his family on the proceeds of that.' He explains how he went out three weeks ago, and paid out an unprecedented 3,000 yards of net, only to find that the little he had caught had been ripped out of the net by seals. 'There's nothing else out there for them,' he says. 'They're starving.'

I know it's coming, and come it does. 'If this had happened anywhere other than the North East, they would have sat up and taken a proper interest in it. But, if it's round here, they don't care.' He expresses his sentiments more gently than most of us might.

After a while his daughter, Sarah, comes over from where she lives next door. 'Here comes my crew,' he says, with a proud smile. She is the deputy head of a primary school and has come to rely on the downtime that his evening and weekend fishing trips provide

---

[15] Video dated 31 May 2022, just over eight months after the initial die-off.

[16] Dr Caldwell's view was that the loss of shellfish stock at James Cole's Whitby was around 20 per cent, whereas in Hartlepool it was nearer 95 per cent.

[17] The medium-sized crab with red eyes that you sometimes happen upon on the beach.

as a release valve for the stress of her job. 'It's an essential part of my mental health,' she says. 'Or was.' For a second or two, the memory of her words dangle in the late morning silence, more powerful than if she had said more.

'On hot nights, we used to paddle our feet in the water down there.' She points out of the window in the direction of the harbour. 'Not now we don't. I don't even bring the kids from school walking here any more.'

I ask them about compensation, just as I had asked James Cole back in Whitby. 'If they said to me,' says Stan with slow deliberation, making sure that I am taking in every word, 'you can have £100,000 or your ecosystem back, I'd take the ecosystem every time.' As it is, they offered nothing, so he didn't have to make the choice.

A few weeks later, I find myself sitting in the London office of DEFRA's Chief Scientific Adviser, Professor Gideon Henderson, who has agreed to discuss with me the robustness of the report of the committee of experts that he chaired.

While he accepts that there is much that was highly unusual about the die-off, not least in its scale and in the strange twitching of the crabs, and things that he might change in hindsight, he does not believe that it was unique, nor that his report was in any significant way wrong. He is clearly surprised that the independence of the members of the committee was called into question and is at pains to underscore what he believes are the two central points: that the die-off was highly unlikely to have been anything to do with the dredging, and vanishingly unlikely to have been pyridine. He takes great trouble to ensure that I am understanding what he is saying.

'There was about a thousand times too little pyridine lying in the estuary sediment to have created a die-off of this scale in the first place, multiplied by another thousand times by the distance it would have travelled to kill a crab in, say, Whitby.' He smiles and adds, in case I hadn't worked it out, 'That's a million.' When I suggest that pyridine could have been a local factor in the Tees Estuary itself, he remains adamant that there wasn't enough of it even to have *that* effect.

I ask him whether it was normal for a pathogen to have jumped species, from crab to lobster to barnacle to the seaweed on Stan Rennie's harbour wall, as this one seemed to, and whether any research had been done on Sally Bunce's seals.

'Certainly, it is unusual to see many species dying,' he agrees. 'But the overwhelming majority were crabs, and there may be other possible and plausible reasons for those other deaths: something emerges in the food chain in the gaps left by the crabs, for example, and eats the barnacles at a greater rate than before. Or local people are just more attuned to seeing dead creatures on the shoreline, even if they had died naturally.' The report didn't look at mammals, so, like all good scientists, he is anxious not to answer a question that wasn't originally even asked.

'I wouldn't know,' he says.

When I ask him about the future, and about whether the precautionary principle should be applied in the case of further work on the Teesport,[18] he wants to make sure that I am not backtracking to the pyridine issue before he answers.

'Are we going to be more prepared next time? Yes, I think we are, particularly in our definition of pathogens and our understanding of chemicals like pyridine. Are the current dredging regulations being applied properly? I've no reason to think that they aren't. Are those regulations tough enough? I think so, but they need to be reviewed on an ongoing basis.' Many of the people I spoke to

---

[18] 'An approach to risk management, where, if it is possible that a given policy or action might cause harm to the public or the environment and if there is still no scientific agreement on the issue, the policy or action in question should not be carried out.' EU Lex Briefing Note

on the ground might have replied, if they had been present, that no regulation is tough enough if the responsible agencies are too hollowed out by cutbacks to enforce them properly. Abuse, they would say, is rife.

Government scientists are in a tricky position, in that they are servants to two masters, the independent science that they have been hired for and the department that employs them. Professor Henderson is manifestly not the consummate bureaucrat; he is more anxious that I understand the science he is trying to explain than that I agree with it, and extremely aware of the pain and hardship the die-offs have brought to an already hard-pressed fishing community. He is a human who happens to be a scientist, as they all are, rather than the other way around.

He, much like the then MP for Scarborough and Whitby from whose offices I have just come, sees no villain in all this, and no smoking gun; instead, I sense that they both – Henderson as a scientist and Sir Robert Goodwill as a farmer and politician – simply accept it as yet another demonstration of the power of nature to deliver unwelcome surprises which then have to be dealt with.

Both are at pains to point out that stocks are improving once again.

'We have to move on,' says Goodwill.

A few miles further east, and a few hours later, I ask Dr Eleanor Adamson (the Fishmongers' academic who made the money available for a number of pieces of research on the die-off, including Caldwell's) whether she thinks my conclusion on the precautionary principle is just some unrealistic dream. She is forthright in her response.

'Not at all,' she says. 'At the very least, initial results suggested a cautious approach. And I would say that there is a clear need for a baseline survey to check for metals and toxins in the sediment.'

When I press her as to whether she thought that there would be one, her silence made the point even more eloquently than words. Happily, she has sufficient faith in Caldwell to have secured for him a small new grant to check for levels of heavy metals in locally caught seafood, as well as for postgraduates at another northern university to explore coastal pollution in the estuary and wider Durham coast.

As I head back up the hill to Fleet Street and the train home, it strikes me as faintly counter-intuitive that the most stalwart backers of those 'forgotten' fishermen in the North East are the people who are privileged to work in this rich and gilded livery company in the comfortable heart of the City of London.

'They were bloody life-savers for us!' Stan Rennie said to me.

For me, the old principle of Occam's razor,[19] which states that the explanation of what has just happened is most likely to be the one reliant on the fewest new assumptions, would suggest that a mass die-off on the edges of the most polluted estuary in Europe was probably in some way connected to the effects of that pollution.

The measure of who was right, and who wrong, will only be seen in the pattern of recurrence as further capital dredging occurs. In future years, I predict that there will be a number of whistle-blowing events that will cast serious doubt on whether the current dredging safeguards are fit for purpose, or properly policed. After all, in a leaked document obtained by Channel 4, CEFAS (the relevant licensing authority) as much as admitted that not all toxic material could or would be stopped from being

---

[19] After William of Ockham, fourteenth-century Franciscan friar and philosopher. You can also call it 'ontological parsimony', if that makes you feel any better.

sea-dumped.[20] Why wouldn't a national government want as much information as they could have on that before leaping into the unknown? The experience of Minamata Disease in Japan in the 1950s, when thousands of people were poisoned by mercury from a local factory that they ingested via seafood that had been caught in the bay, would seem a pretty good, if extreme, argument for caution.

Later on, I find myself thinking back to that conversation with John Aitchison at Crinan all those months ago, when he talked about the precautionary principle. Then, it was in relation to salmon farming; now it is about the biological health of the Tees Estuary in the face of an enormous civil engineering project along the waterfront of an old and contaminated shoreline. In neither instance does it seem to be applied. For good reasons as well as political expediency, the government are determined to create a freeport, with all the jobs and trade that go with it, but without the certainty that it is not going to cause damage to the wider ecosystem. If it was the pathogen that the experts concluded it was, it seems surprising how little subsequent work has been done on where it originated and whether it might come again. You might think that agencies of the government of the sixth largest economy on earth, having definitively come up with two completely different possibly causes for the die-off, might, in the face of a third, have the humility to pause and allow a comprehensive survey of what chemicals lie where in the river bed prior to the main dredging work of the freeport, just to make sure that they weren't about to do something catastrophically damaging for future generations.

---

[20] CEFAS: Centre for Environment Fisheries and Aquaculture Science. The relevant passage reads: 'Considering the results in both the context of the Tees and in comparison to previous years' data, the PAH results presented for this review do not preclude material from continued disposal at sea.'

To think that, of course, would be naïve. That's not how we do things round here, especially when private investors' money is involved, something that has only happened here, by the way, because the council spent the full £276 million clear-up money that the government had given them long before completion, and private money was the only option short of mothballing the site.[21] Private money, as we have all come to learn, is not endowed with much patience. In Mark Cocker's important campaigning book on the state of British nature, *Our Place*, he writes in detail about the creation of Cow Green Reservoir in 1971 after a bitter and hard-fought campaign to save the fell, and with it the old and unique post-glacial flora of Widdybank Fell. But when the ICI chemical plant in Billingham, further down the river, needed more water for its industrial processes, the protests fell on deaf ears and more water was what it eventually got. Almost predictably, it didn't take more than a few years for the plant to start winding down and finally shut completely. It is the same with the HS2 project, where nearly 700 Local Wildlife Sites covering 9,696 hectares, and 108 ancient woodlands are at risk of being significantly affected or destroyed by a vanity trainline for which fewer and fewer of us have any appetite.[22] It will be the same with the proposed Wash Barrier. Granted, protocols have been set up since the 2021 die-off to rapidly take, secure and test samples in the event of a future episode, although the ability of government agencies that have been hollowed out by years of austerity to deliver them effectively must be highly questionable. Certainly, according to the staff members I spoke to, it is.

But there is a deeper, fundamental problem. Every single time, every single project like this, human exceptionalism seems to

---

[21] You missed out on a good deal if you didn't get involved. Private investors picked it up at £1 per acre, or £110.35, plus VAT, in all.

[22] *What's the Damage. Wildlife Trusts report on why the HS2 will cost nature too much.* Jan 2020

triumph, because it is simply easier for politicians to give in to short-term human demands than the long-term needs of our wider nature. More often than not, it is the siren sounds of job creation that leads the battle for justification. And more often than not, those jobs turn out to be temporary, like lanterns on the moor that recede into the mist. Meanwhile, and bit by bit, we continue to play out the tragedy of the commons and to degrade what we have.

So many of the issues that people like me write about are down to the great and increasing number of humans on the planet, about whom we can only be observers, and change nothing. Yet, even with a huge and growing population, we can live different, fairer, better planned and less wasteful lives if we choose to. It's just profoundly depressing that, given time, humility and patience, we could probably have had both the jobs and the natural bounty.

Meanwhile, however much or little that particular chemical had to do with the die-off, pyridine is just another tip of that grubby iceberg towards which our ship is quietly sailing.

Fish and Chips

Seahouses

'It's all in the oil,' says the man from Seahouses.

The further north I travel, the more my principal currency has become fish and chips. For that is how I often pay people for their many kindnesses and also how I reward myself for hard days.

Contrary to what most of us are brought up to believe, fish and chips are about as British as paella or sushi. The 'fish' element is supposed to have been imported by Sephardic Jews in the late seventeenth century, who used the preserving qualities of the batter to cook their fish on a Friday night and then consume it on the Saturday, so as not to break religious laws that commanded them to cook nothing on the Sabbath. Right back in the late 1700s, a cookbook refers to 'the Jews' way of preserving all sorts of fish'. The origins of the 'chip' part are not quite so clear, but almost certainly the idea was brought to England by Dutch or Belgian immigrants, only to find itself married up to the fish in due course. Whatever the origins, by the early 1900s there were around 25,000 fish and chip shops in Britain,[23] and it was an industry considered so important to national morale that it was protected from rationing in both world wars. Since then, the humble 'chippy', with its dodgy health benefits, has survived just about every national food fad and fashion.

I know exactly where the habit came from in my own case. At the age when my friends and I were drinking underage in any local pub that would serve us, fish and chips came as a daring rebuke to home-cooked food and unaffordable steak, a tiny act of rebellion

---

[23] These days, there are still 10,500 fish and chip shops in the UK, turning over £1.2 billion a year and using up more than 10 per cent of the entire domestic potato crop and 30 per cent of its white fish. To settle any arguments that may be brewing, cod outsells haddock by a factor of around two to one (National Federation of Fish Friers).

against our privileged diet. We would start each evening out at the Tasty Plaice in Petworth, the 850cc engine of the Mini revving on the double yellow lines outside, with cod and chips served with salt and vinegar, whether you asked for it or not, and wrapped in half of yesterday's copy of the *Daily Express*. Occasionally, one of us would break ranks and buy a battered sausage with curry sauce, but we were all still far too self-consciously middle-class to order mushy peas. Mushy peas were, like lager tops and *Coronation Street*, beyond the pale.

For obvious reasons, to me at least, fish and chips started to make more sense when they were within sight and sound of the sea than they did 30 miles inland so, for many years, I dropped the habit. These days, even if the fish was caught a thousand miles away, I find I can mentally reduce the food miles involved to the yardage between the fishing boat down there in the harbour and my tummy, even though the boat is full of lobster creels and has never so much as landed a single haddock. The dish is a coastal delight.

I am wary of culinary generalisations, as a rule, so it may just have been serendipity that every fish meal I have had north of the River Humber has been finer than any I enjoyed to its south. From Withernsea to Lindisfarne, they have just got better and better, offering new glimpses of sophistication and reigniting in me a taste that I had lost more than three decades before. It seems that the less the particular town has to prove about its tough maritime heritage, the better its carry-outs. On one particular day, I have haddock and chips in Filey for lunch, and then cod and chips in Scarborough for my dinner, and even then I will be ready for more in Whitby a couple of nights later. I am not proud of this gluttony, to have consumed far more than my fair share of the ocean's bounty, but in time I will make it right.

Some days I will be in a cod mood, some days haddock. Haddock, with its stronger flavours, is for harbour-side bus shelters on rainy nights, waterproofs and phone calls home. Cod, tender and succulent, is for the end of the pier, feet dangling reflectively over the edge and herring gulls swirling around in the reddening sky above. However much tartare sauce I am given, it is never enough, and, sadly, mushy peas are still off my personal menu. In Hornsea, the lady serving me simply cannot believe that I don't want peas and serves me a double portion anyway.

Then suddenly, a few short months after I rediscovered this peerless part of our national food heritage, I walk away again. One last, gorgeous and memorable portion of haddock in Seahouses, almost within sight of the Scottish border, and I know that I have come to the end of that road of delight. The time is right and there will be more another day.

Delights have been relatively few and far between in recent days, so the fish and chips have become much more to me than just a quick meal. Which is surprisingly often how delight works.

BLACK-EYED GANNETS AND THE POWER OF HOPE: NORTHUMBERLAND 13

Late March/Early April

> 'If we throw mother nature out of the window, she comes back in the door with a pitchfork'
>
> Masanobu Fukuoka, Japanese farmer and philosopher

In an old Ministry of Defence hut in some dunes south of Seahouses, James Porteous is trying to give me a flavour of how it was to be on a front line.

James is a young area ranger for the National Trust, covering a dozen miles of coastline just south of the Farne Islands. He came up here in 2018 as a maths graduate who, having achieved the degree and a pile of debt, made the financially inconvenient discovery that all he wanted to do was be outside, and preferably with birds. It's a familiar story: nature's infrastructure round here is utterly reliant on people making similar decisions. After a couple of seasons on the Farnes, he has now graduated to responsibility for the whole section of coast to the south, which includes the largest mainland colony of Arctic tern in Britain, at Long Nanny.

'On a good year we might have about 2,800 pairs,' he says. 'On a bad one, around 1,600. The average would be around 2,000, and no one really understands what influences those numbers from year to year.'

He had known that bird flu was coming. In 2022, the nearby Farne Islands were heavily impacted and lost about 7,000 birds. He was hyper-vigilant that year, but in the end his Long Nanny terns got away relatively lightly, maybe a loss of 50 or 60 birds.

'In 2023, we were on track for a bumper year,' he continues. 'There was an average of two chicks per pair and, for a time, no sign that there was anything wrong. Then we started to notice a few chicks dying, not an unusual occurrence in a busy colony, and we initially put it down to heat exhaustion from the really hot spell we were experiencing.

'Then, one evening, I noticed 50 dead chicks – just chicks – in a small area. That was unheard of and could only have been one thing. I went back to the shed, suited up in full PPE, and then came back to the colony to collect the bodies. At that point, there weren't too many adults involved, but the chicks were taking a hammering.'

And on it went, evening after evening, until they had recovered the bodies of 1,200 chicks and a handful of adults. It was a story repeated all around the northern coasts and left the team wondering how, or even if, the long-term population would ever recover. At times, it all seemed like a glimpse into hell.

To understand from where this virus sprang, you need to go back to 1996 and a Guangzhou goose, resident in some overcrowded southern Chinese farm. That's where it started.

Bird flu has always been around. It circulates naturally in wild ducks and geese and doesn't seem to do them much harm. It is the recently evolved versions that are problematic, often deriving from crowded poultry farms with little or no biosecurity. Indeed, no scientist with whom I subsequently discussed the disease disagreed with my suggestion that it would not have happened, at least not the way it did, were it not for those crowded poultry units.[1] The twist is that, like any pathogenic disease, it will spread quicker through crowded spaces (such as poultry sheds), which can in farming terms be isolated, than it will through widespread populations, but that one of its main engines of spread (along with the movement of poultry by truck and train), wild bird migration, can't be stopped at all, and neither can wild birds be stopped from coming into contact with unhoused farmed birds. Which all means that it keeps coming back, however much you might lock domesticated birds inside their sheds.

Since the mid-2000s, the disease has been spread around the world by the movement of migrating birds, especially wildfowl: a mass die-off of bar-headed geese in western China in 1997 was simply one early physical confirmation of its relentless global journey. Along the key migration flyways it has ebbed and flowed; whether it was a barnacle goose commuting between Svalbard and the Solway Firth, or a red-listed roseate tern flying in from West Africa, there has always been the chance that one or more birds will be carrying the virus with them, and into their crowded breeding sites or their dense wintering flocks. Because seabirds tend to be the most colonial breeders of all, it is on our coast and offshore islands where it has hit hardest. And because seabirds have what are known as 'slow life histories', meaning long maturity times and slow breeding rates, the loss of even one adult can be a real blow. Its 2022 iteration, which is the one I have been seeing my entire journey, has been found in no fewer than 60 countries, from Nepal to Niger, and Britain to Bolivia,[2] and has spread to mammals as varied as bears, foxes, seals and skunks. More worryingly for scientists, what with

---

[1] Actually, that's untrue. A government scientist told me that I was jumping to conclusions.
[2] World Organisation for Animal Health. Paris.

their close proximity to their human keepers, it is now affecting dairy cattle in the USA.

Much of disease control, as we all discovered in 2020, is about preventing movement, something that you are unlikely to be able to do, say, with respect to a wild barnacle goose flying to the Dumfries coast from Svalbard some windy October day, or a guillemot coming in from the Atlantic to breed on its crowded Skomer Island ledge. And, as I saw in Norfolk a couple of months back, a good third of Britain's bird species commutes onto and off our island shores twice a year. No one will ever know how many birds have died,[3] as so many of them will have done so out at sea or deep in the woods, but it is possible that some populations will never recover. Just on that one afternoon on a Cornish beach a few months before during our Lego hunt, Wyl Menmuir and I had seen three dead guillemots and a gannet. And that was long after the worst of the outbreak was supposed to have passed.

If you think back to the larger issues facing the country at the time (the Ukraine war, cost of living crisis, human migration and the tragicomedy of the 45-day Truss 'administration'), you can dimly understand how this unfolding disaster attracted as little of the government's attention as it did. And yet for countless people like me, into whose lives birds inexplicably enter as a currency of uncomplicated joy, it came as nothing less than a daily bereavement. We might not have seen it every day with our own eyes, but its inescapable evidence was everywhere: on the beaches, in the chicken farms, on the television. But what we were suffering was simply the peripheral pain of the helpless observer. Out in the front line of our coastal cliffs and islands, where gannets and great skuas were staring down the barrel of a possible local extinction, people

---

[3] Official figures aggregated from devolved governments and nature organisations suggested 50,000 deaths overall between October 2021 and April 2023, certain to be a significant under-exaggeration (*Guardian*, 5 May 2023).

like James could only walk the colonies in full PPE and collect the corpses of birds that they had dedicated their working lives to, and drop them into rubbish bags. The manner of illness and death, as might be expected with a powerful influenza, was particularly grim.

James and I go outside in the last of the daylight to walk in his patch. It is winter, so no birds are breeding here now. Indeed, his terns are 10,000 miles away to the south, under a foreign sun. The ruins of Dunstanburgh Castle stand prominent a little way down the coast, and a pair of teal flash by in the half-light to land on a little pond in front of us. The calm of the scene is in total contrast with the events that he is describing. 'Eventually,' says James, 'the emotional immunity kicks in. A team of us were camping out 24/7 by the colony to try and do our best to protect it, and we just got on with the job in hand. Strangely enough, the worst thing for me after a while was an anxiety about my own health that turned almost to paranoia. What we were doing was so dirty, and so unusual, that it seemed almost inevitable that we would be physically affected by it, even with all the PPE we were wearing. And it seemed both strange and unfair that, whilst seabirds were being hammered, birds like the ubiquitous woodpigeon seemed utterly immune to it.'

In many ways, death is routine for a bird warden. It has always been the natural order of things for numbers to be taken by a seemingly limitless 'suite'[4] of predators (seabird eggs and chicks are part of the daily summer diets of foxes, rats, badgers, otters, hedgehogs, kestrels, gulls and crows), and in 2019 they had lost a large number to avian botulism. But there is a limit beyond which normality descends into the unknown, and that is where James spent the short breeding season of the Arctic tern in the summer of 2023. And not just James. Indirectly, I and a million others lived

---

[4] A curious word, in the circumstances, but one that fits in to the general aim to make things sound milder than they are and sometimes enables the British 'conservation' industry to speak about it as little as possible.

in that dystopian world, too. Friend and distinguished seabird professor and author Tim Birkhead, during an unbroken 53-year period of study since he was an undergraduate, had watched his beloved Skomer Island guillemots recover from around 2,000 (after the oil spills of the 1960s) to 30,000,[5] only to see the colony ravaged by flu.

Like many of the wardens I have spoken to on my travels, James is undramatic about the events and surprisingly optimistic about the future.[6]

'We shouldn't be too shocked and surprised by it,' he says, as we get back to the hut. 'Disease has been part of any animal's experience since the dawn of time, and this is no different. Immunity should start to kick in, given time: it already seems to be doing so for some of the gannets that they are studying. I guess that these are all natural processes that are being knocked out of balance by the way we live. That's all.'

As for those fish back in Langstone Harbour ingesting microplastics and untreated sewage, so with James's terns, both victims of the choices we make. Thus, he is only articulating a trope of my whole journey.

A few days later, and nine months after I left it, I'm back on a Scottish coastline.

Fifty or so miles to the north-west of Seahouses, I am walking a section of the Forth coastline with Dr Sarah Burthe, an animal

---

[5] As with all numbers in nature, context is everything. Twenty years before Birkhead started studying them, there were around 100,000 Guillemots on Skomer during the breeding season. Thus, the 2023 figure of 30,000 was only a third of the way back.

[6] With some reason, it turns out. When I spoke to him six months later, the 2024 colony was nesting at about 80 per cent of its previous strength, an indication that the 2023 chicks had taken the brunt of the fatalities.

population ecologist at Edinburgh's UK Centre for Ecology and Hydrology. I'm trying to get to the bottom of how hopeful we can be that the worst of bird flu is over, and what might be coming along in its wake. Besides being a scientist, she has the eye of an artist and the swift lope of a competitive ultra-marathon runner, qualities that combine to make the afternoon both highly visual and unusually energetic. The conversation hops around from purslane to violet sea snails and all points in between. By the time we are drinking coffee at the end of our walk in the pretty streets of North Berwick, we are deep into the future of mankind.

'These intensive poultry farms are just giant petri dishes for disease,' she says, when I ask her to confirm my suspicions that bird flu has our own factory-farming fingerprints on it. 'That factor is not being taken seriously or even talked about. And it's not just flu: we are not discussing other nasty examples of disease that are emerging in humans due to links between high-intensity farming and interactions with wildlife, for example Nipah Virus.' Later that evening, I google it, and rather wish I hadn't.[7]

It seems an incongruous subject to be discussing in the bright Lothian sunshine, as we pass exquisitely manicured golf courses, extravagant seaside homes and the joyful blues, greens and yellows of a late spring afternoon. Everything is beautiful, apart from what we are discussing.

'The media covers only what immediately affects its human marketplace: shutting up free range chickens, farm slaughters or what to do when you find a dead gannet on the beach[8] – but it is strangely silent on the wider ramifications. Like that the disease is already affecting Adélie penguins in the Antarctic and the fact that

---

[7] Nipah is a zoonotic virus whose reservoir is in fruit bats, and which kills 40–75 per cent of people who get it.

[8] Call the DEFRA helpline (03459 33 55 77) and then get on with the rest of your life as you await a reply.

it has killed at least 24,000 sea lions along South American coasts in less than a year, and now seems to be coming for elephant seals and mink farms. I mean, sea lions are mammals, aren't they? Like we are?' In the rhetorical question lies a deeply troubling sequence of logic that even a non-scientist like me can piece together.

We stop and look out towards Bass Rock, glowing golden white in the lowering sun with the shit from the 50,000 or so gannet pairs who breed there. Research indicates that the Rock lost about 30 per cent of its numbers in 2022/23, due to the disease, an especially significant statistic as Bass Rock is the biggest gannetry in the UK, which itself holds 55 per cent of the world's population.[9] But there is a quiet optimism among people that I have spoken to in the last few days that the worst is now past.

'Yes, it seems that birds slowly develop immunity,' she answers, when I ask Sarah if she has confidence that things will improve. 'Some quicker than others. At least that seems to be the case with gannets. I'm cautiously optimistic that perhaps the seabirds, as we have seen for the gannets, might evolve to get infected with flu but survive the infection.' Scientists are at their most excited, in my experience, when they are talking about things that no one yet fully understands, and her eyes gleam when she tells me about the phenomenon of gannets' iris colours changing from blue to black when they have recovered from flu.

'It will certainly make it easier to check on them,' she smiles.

I notice that two recurring themes have punctuated her observations throughout our three-hour walk.

'We need to have much better biosecurity around intensive farms and learn the lesson that it is far easier to prevent a pathogenic flu outbreak at source than to contain it once it is in the wild bird population. We need to seriously consider how we farm, and we need much, much better surveillance, especially of live birds. At the moment, all the surveillance focuses on dead birds; but to really

---

[9] The Wildlife Trusts.

understand flu, and to see whether species are coming into contact with it, or showing signs of recovering from it, you need to sample live birds and look for antibodies. And anyway, without live surveillance, you won't detect new pathogens that are starting to appear. Right now, there's plenty of theoretical science, but not much ecology.' In an odd way, we both agree, the experience of Covid-19 actually left society less worried about pandemics than the other way around, less worried than it should be, maybe. After all, more than 99.9 per cent of us survived it. People in her line of work are acutely aware that this sort of luck can run out almost overnight.

'That's why I would like to see us taking all this a lot more seriously,' she says. 'And publicly.'

What concerns her even more, though, is the manmade environmental changes that have ensured that new viruses arrive into an already weakened population.

'Just take two examples,' she says. 'Warming seas and overfishing. Combine those and you'll find that much of the seabirds' diet has moved north and out of range. And then add a third, increased storm events. Storms stop birds fishing, especially cormorants and shags who need to dry their feathers before they go back out to sea.'

The walk done, we stare out over the Forth Estuary in companionable silence. Eventually, I ask her what the large low island out in the middle is.

'The Isle of May,' she says. 'You ought to go out there. Reignite the joy of your inner seabird.' She tells me that her partner has been working there for the last twenty summers.

Having booked the boat passage for the next day, I wander casually over the Scottish Seabird Centre bookshop to check that they are selling *Shearwater*. Because, whether we admit it or not, that's what writers do. After all, if you can't sell a book about Scottish shearwaters in the Scottish Seabird Centre bookshop, then where can you sell it?

'No,' says the lady at the till. 'We haven't come across that one. But there's lots of others for you to choose from.'

The following afternoon, I am sitting inside a large wooden crate strapped down to the very edge of a 50-metre cliff, staring out at a rock face opposite and slightly below. The crate has two makeshift windows within its walls, which provide a vista onto part of the avian city beyond.

'It's my office,' says field manager for the Isle of May, Mark Newell.

The level of detail to which natural history fieldworkers and academics go to in their research and surveying is quite extraordinary to a layman like me, both in terms of the timescales (often half a century of unbroken research) and the granularity involved. It is the aggregation of all these surveys, species by species, reserve by reserve, that allows us to know with great accuracy what is happening to our seabird populations, and why, and what if anything, we can do about it.

'Take a look at this map,' he says, showing me a laminated A4 sheet, on both sides of which are drawn intricate plans of each individual kittiwake and guillemot nest site, numbered and complete with an up-to-date status. 'So that pair on the right-hand edge of the ledge down there, number 223 on the map, have returned, and I can now check them in for the season.' He circles the number on the map with a red pen. 'There's over 5,000 pairs on the islands, down from 8,000 a couple of decades ago, and then in a week or two they will be joined by 15,000 pairs of guillemots,[10] who we also monitor.'

---

[10] This figure subsequently dropped to 13,000 pairs in the poor 2024 breeding season. Their breeding success rate was 65 per cent below the long-term average of 72 per cent, but well above the awful years of the mid-2000s.

I remember back to that Sandwood silence on the first day of my journey, now nearly a year ago, and to what I have learned along the way since. I have learned, for example, that seabird numbers are a mixed picture, even if the overall direction is down. The authoritative Joint Nature Conservation Committee (JNCC)[11] report on seabird numbers of November 2023 shows that eleven of the 21 studied species are declining, five are static and five increasing. Basically, good news if you happen to be a Mediterranean gull or a razorbill, bad news if you are a Leach's storm petrel, and business as usual if you are a cormorant.

But numbers are only a bit of the story, the dull frame around the glorious landscape painting in the gallery. They can give you the science, but cannot articulate the comfort that we, as humans, have been given the ability to find in nature, nor the grief when things go wrong. They are also the hints in braille to a leadership whose eyes are often determinedly fixed shut. If everyone who strove to be in a position of political power over the environment had to sit in this little crate until their breath caught in their chest and their eyes moistened with the fragile glory of it all, we might get somewhere.

'You can't help it, really,' he says, when I ask him if he gets emotionally involved in his work. 'When you have held a bird in your hands, ringed it with care, seen it off at the end of the season and welcomed it back the following spring. How can you possibly not mind when there is an empty space one year when it fails to come back? But, equally, you get used to it, and anyone who has been involved in nature for more than a year or two knows that death is just as much part of the circle as new life.'

May will be my last island on this journey, so its treasures are all the sweeter. As my afternoon draws to a close, I sit on a rock for a while and look out at the vertical metropolis of bird activity around

---

[11] The statutory adviser to the government and devolved administrations on UK and international nature conservation.

me and just lose myself in awe at the joy of it all. Stiff-winged gannets commute below me; the season's early guillemots bustle up to their little ledges on the cliffs; a raft of eiders rise up and down on the swell far below. To the south and north, the East Lothian and Fife shores sparkle in the late-afternoon sunshine, and the sea shimmers in its exaggerated, impressionist beauty.

These precious moments in time perhaps explain in part why so many of us try to make our lives and our living out of nature, indeed why I set out on this coastal adventure in the first place.

My journey is nearly done, but I have one more appointment to keep.

SALAR: NORTH-EAST SCOTLAND

14

April

> 'To be truly radical is to make hope possible rather than despair convincing.'
>
> Raymond Williams

I haven't picked up a fishing rod in many decades.

Last time I was here, it was 40 years ago, and no one was really talking about the possible extinction of a fish, let alone the wild Atlantic salmon. They are now. Particularly the man I am with.

Andrew Graham-Stewart has come a long way in his 70 years, not least from his time as the manager of a German psychedelic rock band in the 1970s,[1] to the passionate campaigner for wild fish in general, and the Atlantic salmon in particular, that he is today. He has agreed to walk with me up the bank of the only river on which I have ever fished with any degree of commitment, Sutherland's Brora, to help me understand what has happened and

---

[1] Tangerine Dream, pioneers in the *kosmische musik* bit of the German musical scene. As if you didn't know.

what, if anything, can be done about it. He bears his unshakable love for the Atlantic salmon like a well-worn sweater. Many of them do the same round here.

I met him months before on a Zoom call to discuss farmed salmon and liked him instantly because he told me I was arguing against my own logic in my conclusion, which I was. 'Come and walk a river with me in the spring,' he said.

Few things better illustrate the boundless connectivity between sea and river, humans and nature and past and present so much as the wild Atlantic salmon.[2] And few creatures exemplify more powerfully the past effects and future challenges of our stewardship of the waters around and within our islands than that magnificent predatory migrant. Just as my story began with a tank full of farmed salmon off Scotland's north-west coast, so it draws towards its close with a remnant population of their wild cousins in an east coast Sutherland river. Farmed and wild salmon are the same species, but only in the sense that a bloated beef cow in some feedlot is related to the majestic wild aurochs. When I stared into one of the pens at that Loch Duart salmon farm in Badcall Bay the previous summer, it had been extremely hard to equate the closed and circular life of the captive fish with the expansive migrations of its wild cousin.

'It is interesting,' he says, 'how the rise of the farmed salmon has inversely mirrored the relentless decline of the wild one.' It is, and it has. Correlation is not causation, I know, but the two opposing lines are too closely related for the penned fish not to be among the prime suspects. That the decline of wild salmon numbers on the east coast (where there are no fish farms) is as bad as that on the west is often used by supporters of the farms as evidence that the two aren't connected, but it overlooks the fact that all those fish migrating to

---

[2] *Salmo salar.*

and from the east coast rivers generally have to pass by the Orkney farms on the Pentland Firth.

As we walk upriver, he tells me about his wild fish. About how theirs is a river-sea-river story, one of beauty, complexity and staggering navigational precision, a rugged traveller's tale from the northern depths. How it begins with a pea-sized egg buried in an upriver gravel bed, only around one in two or three thousand of which will ever make it all the way through the various stages of life,[3] and closes with the salmon being drawn back to that same river by a mixture of magnetic field, pheromones and the river's very own chemical smell. Of how it is a story of survival against all odds. Just to watch a salmon repeatedly trying to fire itself up a weir or waterfall is to see an instinctive determination that borders on our own interpretation of madness. And it is a story of a supremely adapted cold water carnivore, the very thought of whose current vulnerability seems almost irrational for us to be holding.

'It's a big responsibility on us all not to lose it forever,' he says.

He tells me how the thread of the salmon has been woven into our own island story since before we even had a story to tell. Salmon legends and fables have punctuated local folklore wherever there has been river or sea to inspire them. There is evidence that our Neanderthal cousins were eating them a quarter of a million years ago, and that the Romans and Gauls were constructing ponds to better hold them before Julius Caesar ever set foot in Britannia. We have fished for it out at sea and netted it, poached it and trapped it in wattle cages. With all the glorious irrationality of our own inventive species, we have cast half-hopeful flies over it and spent comical amounts of money doing so. The pursuit of a salmon by means of casting somewhere near it a fly that

---

[3] Egg becomes alevin becomes fry becomes parr becomes sea-going smolt, and only then does it become salmon.

it certainly doesn't recognise at a time when it is not even programmed to eat is in glorious defiance of all common sense and previous invention, an ironic nod to a forbearance that our species otherwise abandoned centuries ago. If our island's natural history could be written, you can be sure that the salmon would be writ large throughout its pages.

And yet, in December 2023, the conservation status of the Atlantic salmon moved from 'vulnerable' to 'endangered', meaning that the species is now considered at risk of extinction.[4] Its journey from such extraordinary abundance that employers were once prevented from serving it to apprentices more than once a week to the foothills of extinction serves simultaneously as a metaphor for our times and an illustration of the extraordinary complexity of ecosystems.[5] The ingredients list of the cocktail that has brought it to this point is a numerous and far-ranging one: historic overfishing in the North Atlantic, extreme weather events, shortage of food, warming sea and river temperatures, dredging, acidification and other pollution, forestry, hydro, dams, weirs, sheep dip, the effects of former netting, poaching, angling, disease, the end of seal controls and, as we saw back in Chapter One, infections from farmed fish. There are more, but all these have one thing in common: the footprints leading inexorably towards them are our own. We can argue all we like about the relative details, but this one is on us. And, like everything else in our scramble to save the planet, the answers are far from straightforward: riverside trees absorb carbon and lower water temperatures, but needles from conifers acidify the water; seals eat salmon, but no one wants to see a culled seal; more hydro power means less fossil-fuel emissions, but three pumped storage hydro facilities on Loch Ness will eventually lead to fewer fish, as it raises and lowers the water

---

[4] IUCN Red list of threatened species, December 2023.
[5] British Newspaper Archive entry, 11 July 1811.

levels of the loch by up to a metre a day, and changes the entire direction of the flow.[6] It's not easy.

The wild Atlantic salmon is the canary in the coal mine. It is to the river, estuary and coast what the grey partridge is to the arable farmlands and the great crested newt is to a pond, in the sense that, through its own condition, it indicates so much more about the wider conditions around it. And never mind the inherent unfairness behind the image of very best salmon fishing being overwhelmingly a sport for the white, the male and the solvent, which it may well be, its gradual death should matter to us all, just as its future recovery could be a poster child for mending ecosystems everywhere.

Turbines churn away massively on the skyline and the first swallows and sand martins of my summer are swooping down for the insects flying over the small floodplain. I ask Andrew provocatively why the salmon's disappearance really matters that much beyond the fishing community. 'I mean, survival is the exception among species,' I add. 'Not the rule. Something else will fill the gap that it leaves soon enough.'

'Well, it's a vital part of the ocean and river food chains, for a start,' he says. 'And it was, until recently, an important source of food and employment to humans as well, not to mention its cultural significance. Then there is the precedent. If we can't save the salmon, what the hell can we save?' He charts the decline that he has witnessed and accepts that we probably don't even really know what the main cause of it is yet. 'Something seems to be happening to their food source way out in the Atlantic. It's moving, or it's reducing. The young ones are just not surviving and, if they are,

---

[6] One (Foyers) is already there; two (Red John and Loch Kemp) are somewhere in the planning process. The process of pumped storage hydro power basically pushes water uphill during hours of low demand (and therefore cheap) power, and then releases it back down through the turbines during periods of high demand.

they are much smaller. We need to do everything we can to buy them time until we can identify what it is and maybe do something about it.'

I ask him about the burgeoning grey seal population, and he dismisses the issue. 'There's a few specialist seals that wait for salmon by obstacles,' he says. 'But research shows that most seals have no trace of salmon in their stomachs. Besides, are we really going to go back to something as publicly unacceptable as seal culling? No. And nor should we.' Then he adds that, if anything, we should be more concerned about the dolphins.

'What about hatcheries?'

'It's an absolute final straw action,' he says. 'It might put more fish in the river, but they aren't savvy wild salmon. It's a sticking plaster.'

For an hour or so, we clamber our way upriver over fences and across ditches, talking as we go. He is a consultant to Wild Fish, the charity whose Scottish arm he used to run, and tells me that he fails to understand how other pressure groups can truly advocate for the wild salmon when so many are part-funded by a Scottish government utterly reliant on any findings not coming down against farmed salmon. In the long cocktail menu of problems for the salmon, he focuses my attention relentlessly on the one he believes that we can do something about, the infective influence of those farms.

'So is that how we buy them time?' I ask.

'Fixing that is the one sure thing we can do,' he says, as we sit on a rock looking down at what he tells me is one of the best spring pools in Scotland. For a while, we watch in silence as a fisherman works his rather half-hearted way down the opposite bank to where we are. His casting is competent, but he looks as if he would rather be back in the lodge eating his lunch, or at least talking to someone.

As we walk back down the road that leads to the village, I ask him, as I have asked the scores of other experts that I have walked with and whose generosity of time and advice I have relied on, whether he is optimistic, in his case about the wild salmon.

'Yes,' he says. 'Salmon have got a chance. And if it turns out that they don't, it won't be for the want of effort by groups like ours, and others. It will be because we were too busy screwing up everything else on the planet to notice, and to mind, about one amazing migrant.'

Sometimes, we both agree, you believe in survival simply because you cannot bear to imagine extinction.

EPILOGUE: DUNNET HEAD

May

It is an acid yellow view that greets the last morning of my journey, and it is a world of birds.

The road to Dunnet Head,[1] when I set out on it, is bordered by gorse, broom, dandelions and celandine. The air is alive with the chatter of larks in the sunshine and the bubbling call of breeding curlew above old hay meadows. To my left, a short-eared owl plots the tracery of its whimsical flight path over a wet field. Somewhere out of sight on the other side of the road, I can hear snatches of a chittering oystercatcher and the mew of a buzzard. I bathe my feet in the last freshwater loch in Scotland in the company of a pair of indignant greylag geese; a willow warbler sings its heart out in a gorse bush a few yards from the end. For a journey that had started in bird silence, it is certainly making up for it today.

---

[1] Dunnet Head lies west and, significantly, 2.3 miles further north of the much less attractive John o' Groats. It has always struck me as strange that people walking or cycling all the way up from Land's End would not want to put in the last few miles to end up in the right place.

The end of long journeys, even broken journeys like mine, is always a tangle of mixed emotions. On the one hand, you have secured that most human of needs, the satisfaction of achieving what you set out to do;[2] life can return to normal, whatever that is. You can return home. On the other, you feel grief at the ending of something both enthralling and challenging. The memories will live on, but the precious carelessness of daily adventure, with all its choices and unknowns, stops right here, right now. To walk for days on end with only the sound of gulls above and the restless ocean below has been to see life utterly anew. To have walked so much of the island's entire coastline has been life-changing in more ways than one.[3] It has been to look through a window onto a world of forces infinitely more powerful and permanent than man, and yet it is a world that, at the same time, is surprisingly fragile and changeable.

I find a quiet spot on the cliff edge and sit as close as I dare to the abyss to eat my sandwich, and to think, appreciating for a moment the symmetry in the wheeling fulmars who had seen me off from Cape Wrath a year ago and who seem now to be welcoming me to Dunnet Head. For a moment, I am the northernmost human on mainland Britain. I think that it is raining, but out of a cloudless sky. However, here, where it is never fully calm, thin pulses of sea spray are carried up over the cliff edge, a full hundred metres above where they started, even on calm days. Out there to my north, I can make out the Old Man of Hoy, Scapa Flow and the little island of Swona with its herd of feral cattle. Moving northwards has always felt right to me, which might explain why I have found the second

---

[2] Maslow would disagree. In his hierarchy of needs, he places 'esteem and self-actualization' at number five. What would he know?

[3] The British coastline is around 11,000 miles. I gave up recording cumulative daily distances quite early on, but it probably amounted to between 1,900 and 2,000 miles, which, in my language, is 2.5 pairs of Meindl boots, 70 flat whites or around 17 per cent of the coast.

part of my journey easier than the first. North is an aspiration, a sensation, a gleam. North is where the water is colder and so where much sea life is now headed. But north is now where I stop.

The coast that I have been wandering cannot be described in isolation, as it is neither finite nor discrete; everything is connected to everything else. The sea is indivisible from the coast, which is indivisible from the land that rises up behind it and the rivers that run into it. What happens ten feet below the tideline is no less important than what happens ten feet above it. The effects of the biological death of the River Wye, for example, even though miles upstream from the sea, spill out into the Severn Estuary, just as the wild Atlantic salmon fails to return to its native river because of something that is happening a thousand miles out to sea. That we are a nation of litter louts does not, sadly, remain irrelevant to the coast many miles from where the litter is discarded, because the coast is where freshwater finishes. That's one reason why the coastline matters so much, whether we go there or not. The truth is that what affects the coastline is rarely because of what happens there. Maybe nothing expresses that less eloquently, but more effectively, than all those tons of discarded ghost fishing plastic, or those dead crabs on Redcar Beach.

Think of it, for a moment, in terms of mammoths. As our ancestors became better and better at hunting them down, first in singles, then pairs and then herds, so they brought the species up to and then through the gates of extinction. Suddenly, there were no more mammoths, and a valuable food source had been removed. This, in anthropological terms, is called a 'progress trap'. Many millennia later, much of what is happening on our coastline is simply a consequence of another progress trap, by which the brilliance of our minds and the heights of our material ambitions are combining to make our home slowly uninhabitable. It's not a good look.

In a way, what I have learned is childishly simple, and I probably knew much of it before: our coastline is a miracle of variety, complexity, generosity, beauty and potential joy. Within reason, if more of us went back there and relearnt how to love it, rather than jumping on some aircraft to visit an overheated southern European beach, it might recruit more defenders, who would contribute more to its economy and generate more noise at the abuses that are inflicted on it. There has not been one day, from Durness to Dunnet Head via Dover, that has not captivated me in some way and taught me things about which I had little or no idea. I have found beauty and life everywhere, just as much in the little creeks of the Thames as in the sea cliffs of Sutherland, and above all I have seen the potential for recovery, much of which is trying to get underway even now. By and large, when recovery starts, it happens surprisingly quickly.

Some things you can't change. Tectonic movement, for one. Tides and the underlying rising and sinking of land, or the erosion of alluvial soil left by the last ice age. There are other things that we can change only slightly and at glacial pace, if at all, such as the rising sea temperatures and the stormier weather. But there are also many things that we can change tomorrow if we have a mind to, beginning by our acknowledging the fundamental fragility of our ecosystems. We could start by applying far greater localism to decisions that are made on behalf of coastal habitats and communities, and then apply the full precautionary principle to intensive aquaculture and to major infrastructure developments; we can eat less seafood, food generally, in fact, and with more discrimination as to how it ended up on our plate, and we can waste much less of it; we can urgently fight the war against single-use plastics and disenable businesses from producing and promoting them in the first place, full stop. We can apply real sanctions to the utilities who fill our rivers, lakes and bays with unnecessary sewage, even to the point of imprisoning their leaders if that is what it takes. To do this requires us to elect politicians who care beyond mere virtue-signalling cliches, and for us to be prepared to consume less and to create in our own minds a vision of just how good it could be in 50 years' time, say, if we lived just slightly different lives. There are less outright villains in this story than idiots driven by greed, laziness and ignorance; their numbers are dwarfed by the small army of individuals trying to repair and enhance what we have been given. It's an army whose ranks need to swell all the time, until the effect of the idiots is neutered.

There is a temptation to think that restorative action is purely limited by the funds available in these straitened times. Maybe, but I suspect not. Maybe, instead, it comes down to choices, and to the fact our governments tend to spend money with reckless idiocy: just think what research Andrew could have done on his salmon with the £500,000 spent on a private charter to get Liz Truss to Australia to 'negotiate' that trade deal;[4] or what Philip Price and his colleagues might achieve at Ardfern with the £11 million extra that went on rushing through post-Brexit blue passports;[5] or what Chris Binnie could build in the Bristol Channel with the £3 billion *extra* that the government spent on temporary agency staff for the civil service.[6] Just think how much plastic we could remove from our coastline with the £2.3 billion fine that the UK had to pay to the EU over cheap Chinese imports.[7] My advice is not to buy a word of the 'no money' story. Protecting and enhancing nature is staggeringly good value in comparison to the waste involved in the cack-handed awarding of contracts, messing up of infrastructure projects, fraud and dodgy vanity schemes that you and I have been saddled with. It's not about money; it's about priorities, and our human ecosystem.

---

[4] Sky News, 27 January 2022.
[5] Home Office report, 31 March 2020
[6] *Independent*, 9 October 2022.
[7] *The Times*, 9 Feb 2023.

The dark sea surges away down below me. By chance, I am eating seaweed crisps,[8] which come from a few miles away, and writing my notes on a phone charged by wind power from a turbine on the hill behind my bed and breakfast. Tonight, I will eat Cullen skink made with haddock caught somewhere over the horizon, and drink whisky distilled from barley grown in the protection of Islay's coastal climate. Nature is generous.

It was maybe a need for this generosity that drew me on this adventure in the first place. While I could always have lingered longer, and seen and learned more, I think I saw enough for me to know the coast more usefully, and to understand what it provides us with. Mile for mile, stone by stone, these edgelands and all that they contain remain one of the greatest natural gifts we have.

I can't stay here for ever. Nearby, an unlaundered photographer shows impatient signs of wanting to use the exact spot I am sitting on to film some fulmars, so I stow my pack, stand up and give one last, lingering look across the sea below me. Anyway, I have seven hot miles to go before I can taste that Cullen skink and drink that whisky.

'Thanks, mate,' he says, neutrally, as if I have been sitting on his personal stool in his private bar. His disruptive green camouflage uniform, once designed for jungle warfare operations, makes him look ridiculously conspicuous on the bare cliff top, miles from the nearest tree.

Meanwhile, somewhere far below, and out there deep in the Pentland Firth, a wild Atlantic salmon is completing the long pilgrimage to the river of her birth.

May she and her successors go on doing so for ever.

---

[8] Utterly delicious. Shore lightly salted crisps, made from seaweed harvested a few miles south of Dunnet Head.

ACKNOWLEDGEMENTS

This book is the product of maybe 500 conversations, some day-long, and some lasting just a few minutes, some on lengthy cliff-top walks, some in huts, some necessarily virtual. A feature of all these exchanges, generous gifts from busy people as they were, was the overwhelming feeling from the experts that this is a story worth telling, and their participation was enthusiastic, not reluctant. I hope that in the list below, I have remembered most of them, and that I have done their expertise and their passion justice. Without them all, there would simply be no book. Without them, too, the British coast would be a far poorer place.

Prof Chris Spray (University of Dundee)
Prof Russ Wynn (Wild New Forest)
Prof Tim Guilford (Merton College, Oxford)
Charles Clover (Blue Marine Foundation)
Danny Renton (Sea Wilding)
Philip Price (Sea Wilding)
Richard Smith (Dee Estuary Birds)
Matt Thomas (Ranger: Hildre Island)
Jane Turner (Hoylake)

Sophie Plant (COAST)
Howard Wood (COAST)
John Aitchison
Geoff Heulin (Ilfracombe Lobsters)
Rachel Yates (Surfers against Sewage)
Rob Rattray
Mark Cocker
Fred Stroyan (Fishmongers)
Wyl Menmuir (Cornwall)
Joe Richards (Blue Marine Foundation)
Rupert Hawley (Plantlife)
Claudia Watts (Royal Parks)
Christoph Harwood (Simply Blue Group)
Prof Alastair Driver (Rewilding Britain)
Sara King (Rewilding Britain)
Don O'Driscoll
Dr Bob Earll (Coastal Futures)
Dr John Howe (SAMS)
Prof Roger Falconer (Cardiff University)
Prof Chris Binnie (Tidal Engineering and Environmental Services)
Louise Valducci (Compassion in World Farming)
Dr Leanne Cullen-Unsworth (Project Seagrass)
Dave Sexton (RSPB Isle of Mull)
Joe Wilkins (UK Youth for Nature)
Laura Evans (Marine Wildlife Centre; New Quay)
Maddy de Marchis (Marine Wildlife Centre; New Quay)
Steve Hartley (Marine Wildlife Centre; New Quay)
James Grellier (Exeter Medical School)
Dr Pamela Buchan (Exeter University)
Alexis Perry (Environment Bank)

Alasdair Mitchell (Ocean Plastic Pots)
Tavish Scott (Salmon Scotland)
Andrew Watson (Salmon Scotland)
Hazel Wade (Loch Duart Salmon)
Andrew Wallace (Fishmongers Company)
Pat Holtham (Uist Hedgehog Rescue)
Merlin Hanbury-Tennison (Cabilla Cornwall)
Moray Finch (Isle of Mull and Iona Community Council)
Stuart Gibson (Isle of Mull)
Conor Ryan (Isle of Mull)
Ewan Miles (Ilse of Mull)
Chris Duffett. (Baptist Church)
Louise MacCallum (Blue Marine. Solent Seascape.)
Eric Harris-Scott (Blue Marine Solent Seascape)
John Goodlad (Shetland Islands)
Prof Ivan Haigh (University of Southampton)
Tom Bennett (Foulness Island)
Nick Acheson (Norfolk)
Sarah Farnsworth (Norfolk)
Dom Buscall (Wild Ken Hill)
Jake Fiennes (Holkham)
Hayley Roan (RSPB Norfolk)
Ray Kimber (RSPB Norfolk)
Prof Mike Elliott (University of Hull)
Peter Wadsworth (Beverley)
Karen Nicholson (Easington, East Yorkshire)
Dr Tim Hill
Kevin Edwards. (Friends of Langstone Harbour)
Rob Bailey (Clean Harbours Partnership)
Prof Alex Ford (University of Portsmouth)

Kabir Kaul
Annie Wisbey
Nick Marks (Lyme Regis RNLI)
Rachel Mulrenan (WildFish)
Andrew Graham-Stewart (WildFish)
Rob Hunton (Spurn Observatory)
Mike Welton (Easington Historical)
Hugh and Fiona Fell (Northumberland)
Vicky Portwain (RWE)
Dr Shaun Nicholson (Marine Management Organisation)
Dr Gary Caldwell (Newcastle University)
Alasdair Robertson (Ornithologist: Natural England)
Sir Robert Goodwill MP
Prof Chris Redfern (Newcastle University)
James Cole (Whitby Commercial Fishers' Association)
Stan Rennie (North-east Fishing Collective)
Dr Eleanor Adamson (Fishmongers' Company)
James Porteous (Northumberland National Trust)
Billy Shiel (Seahouses)
Professor Bill Sanderson (Herriot Watt University)
Dr Sarah Burthe (UK Centre for Ecology and Biology)
Sally Bunce
Mark Newell. (Isle of May)
Nigel Pearson
Lord Richard Benyon
Professor Gideon Henderson (DEFRA)
Kevin Brown ('Thames Estuary Man')
Ruth Cromie (WWT Research Fellow)

BIBLIOGRAPHY

(Or, a suggested reading list in preparation for a long coastal journey.)

Rewilding the Sea	Charles Clover
Tickets for the Ark	Rebecca Nesbitt
The Summer Isles	Philip Marsden
Life Between the Tides	Adam Nicolson
Coastlines	Patrick Barkham
Life Changing	Helen Pilates
Tide	Hugh Aldersley-Williams
The Deep	Alex Rogers
Coasting	Jonathan Raban
Cruise of the Nona	Hilaire Belloc
The Salt Roads	John Goodlad
Salmon	Mark Kurlanski
The New Fish	Simen Saetre and Kjetil Ostli
Sea Fever	Meg and Chris Clothier
The Draw of the Sea	Wyl Menmuir
The Brilliant Abyss	Helen Scales
Silver Shoals	Charles Rangely-Williams
Book of Tides	William Thomson
The Value of a Whale	Adrienne Buller

Britain's Habitats	Sophie Lake et al
A Field Guide to Getting Lost	Rebecca Solnit
Wanderlust	Rebecca Solnit
A Philosophy of Walking	Frederic Gros
After They're Gone	Peter Marren
The Waterside Ape.	Peter Rhys-Evans
Otherlands	Thomas Halliday
The Fabled Coast	Sophia Kingshill
The Science of the Ocean	Natural History Museum
An Irish Atlantic Rainforest	Eoghan Daltun
Spirals in Time	Helen Scales
The Meaning of Geese	Nick Acheson
Sea Change	Richard Girling
One Midsummer's Day	Mark Cocker
The Outlaw Ocean	Ian Urbina
Home Waters	David Bowers
The Seaside: England's Love Affair	Madeleine Bunting
Wild Mull	Littlewood and Jones
The Coffin Roads	Ian Bradley
Tide Race	Brenda Chamberlain
Reflections	Mark Avery
The Turning Tide	Jon Gower
Shaping the Wild	David Elias
Water Always Wins	Erica Gies
The Britannias	Alice Albinia
Moderate, Becoming Good Later	Toby Carr and Katie Carr
Bird Migration	Ian Newton
Blue Machine	Helen Czerski
The Flow	Amy-Jane Beer
Sea Bean	Sally Huband
The Way to the Sea	Caroline Crampton
Illuminated by Water	Malachy Tallack
A River Runs through Me	Andrew Douglas-Home
Casting Shadows	Tom Fort

The Old Ways   Robert Macfarlane
The High Seas   Olive Trennernan
The Human Planet   Lewis and Maslin

Mission Zero. The Skidmore Independent Review on Net Zero
COP15 Global Biodiversity Framework Report
Marine Climate Change Impacts Partnership Report Card 2020 (2022?)